James M. Duncan

Clinical Lectures on the Diseases of Women

Delivered in Saint Bartholomew's Hospital

James M. Duncan

Clinical Lectures on the Diseases of Women
Delivered in Saint Bartholomew's Hospital

ISBN/EAN: 9783337239503

Printed in Europe, USA, Canada, Australia, Japan

Cover: Foto ©berggeist007 / pixelio.de

More available books at **www.hansebooks.com**

CLINICAL LECTURES

ON THE

DISEASES OF WOMEN,

DELIVERED IN

SAINT BARTHOLOMEW'S HOSPITAL.

BY J. MATTHEWS DUNCAN,
M.D., LL.D., F.R.S.E., ETC.

PHILADELPHIA:
HENRY C. LEA.
1880.

SHERMAN & CO.,
PRINTERS AND STEREOTYPERS,
PHILADELPHIA.

To

Dr. FORDYCE BARKER,

𝔄 Worthy Representative

OF

AMERICAN OBSTETRICS AND GYNÆCOLOGY.

PREFACE.

THESE Lectures were originally published in the *Medical Times and Gazette* and in the *Medical Examiner*, at the request of the Editors. They are now reproduced in a separate form at the instance of the Publishers. The Lectures that appeared in the *Medical Times and Gazette* are almost word for word as given in the class-room, having been taken by a shorthand writer. Seven were first published in the *Medical Examiner*, from notes taken by Dr. Godson. I would not have brought them out in their present form had I not received suggestions and encouragement from professional brethren, at home, and in France, Germany, and America.

It will be obvious to the reader that the naming of authorities and literary references is avoided almost entirely; and this is done for good reasons. The chapters are Clinical Lectures to Students, and the whole object of the teacher was to increase the acquaintance of his pupils with disease. The teacher had no time for anything however slightly foreign to this purpose. Even if he had had time, the divergence into historical details would, he believes, have detracted from the efficiency of his teaching. It must not be supposed that he attaches little value to authority and to literary detail—quite the contrary. Indeed, he makes much of such matters in his Systematic Lectures, where they find an appropriate place.

He has to thank Dr. Godson for assistance in passing the work through the press.

Finally, he expresses hope that Dr. Fordyce Barker will pardon the liberty he has taken of dedicating the work to him without previously asking his permission.

71 Brook Street, W.,
            November 1st, 1879.

CONTENTS.

CLINICAL LECTURES

ON

DISEASES OF WOMEN.

I.

ON MISSED ABORTION.

MISSED ABORTION is a subject that lies between obstetrics and gynæcology; the cases indeed that I am particularly to dwell upon were brought into "Martha" as gynæcological cases, or cases of diseases of women, more than as obstetrical cases. I do not know any subject better than this for illustrating the value or necessity of extensive knowledge with a view to good diagnosis. If you do not know of a thing, you are quite sure not to suspect it; and, in all cases of difficult diagnosis, if you do not suspect a thing you are almost certain not to find it. This remark is especially true of the subject I have to consider now.

A missed abortion is not a threatened abortion, nor is it an imperfect abortion. A threatened abortion is a very common occurrence. When a woman has a threatened abortion she suffers pain, she has bloody discharge, and the mouth of the womb may be found to open. An abortion may only get the length of being threatened; that is to say, the abortion may be averted and pregnancy may go on healthily, even when you have been able to feel, through the neck of the womb, the ovum as it hangs in the cavity of the body of the uterus. I have known also two cases in which a considerable piece of decidua was separated and discharged without abortion taking place. It is natu-

2

rally expected that, as has been shown to be the case in placenta
prævia and in the separation of decidua in extra-uterine pregnancy,
the detachment of bits should take place near the internal os, where
it would least disturb the ovum.   These are cases of threatened abortion, and among them may be included cases of extreme rarity, of
abortion of one of twins, while the other remains in utero and goes
on in its development.

This abortion of one of twins may be a missed abortion, or the
miscarriage of one of twins may be a missed miscarriage. In that case
the fœtus and its envelopes, instead of getting rolled up into a parcel-
like form, as I shall describe to you, become compressed and squeezed
flat between the uterus and the growing ovum into the condition
which, when extreme, is called fœtus papyraceus.   I show here a
beautiful specimen of fœtus papyraceus, occurring in a case of twins
where there was missed miscarriage.

To recapitulate : In cases of threatened abortion you may have a
discharge of a bit of decidua, you may have the neck of the womb
open to the extent of allowing the finger to pass and to feel the ovum,
and you may have a missed abortion or a missed miscarriage in the
case of twins.

Missed abortion is neither a threatened abortion or miscarriage, nor
an imperfect miscarriage. In order that you may understand an
imperfect miscarriage (of which I have a remarkable instance to describe to you), I must tell you what is a complete or perfect miscarriage.   If the fœtus alone, or the entire ovum alone, comes away, the
woman has miscarried, or aborted, as it may be ; but the coming
away of the ovum does not involve a complete miscarriage ; and an
imperfect miscarriage is often a very disastrous thing.   The ovum
sometimes comes away alone, without any of its uterine or maternal
membranes.   Sometimes the fœtus comes away alone, without even
the ovuline membranes.   Sometimes the ovum comes away, and the
maternal membranes or decidua imperfectly.   Sometimes only a bit
of the placenta is left, as in the case that I am to relate.

Imperfect miscarriage is a dangerous thing, frequently in consequence of the very serious and recurrent bleedings that result from it.
It not very rarely leads to death from mere putrid intoxication or
sapræmia, or from septicæmia, or from pyæmia, just as happens after
delivery at the full time.   This is especially liable to occur if the
miscarriage has come on in consequence of extensive endometritis such
as is found in pregnancies occurring during typhoid fever.   Imperfect

miscarriage is also often disastrous by inducing endometritis, generally purulent endometritis, and this frequently in connection with putrefaction of the parts left behind.

The case of imperfect miscarriage which I am about to read is in every respect remarkable, and illustrates the subject admirably. M. C., aged thirty-eight, married for sixteen years, has had six children, the last two years ago. On March 14th, that is, eight months ago, she miscarried with a three-months' fœtus. The placenta did not come away till three weeks afterwards. Subsequent history shows that the whole placenta did not come away even then. For a fortnight before, and for six weeks after the miscarriage, she had considerable bloody discharges. Since then losses of blood have occurred occasionally. She is feeble and anæmic, but otherwise healthy. Nothing abnormal, except suprapubic dulness, discovered on examination of the hypogastrium. Digital examination per vaginam finds the cervix uteri largely patulous, greatly hypertrophied, but not softened as in pregnancy. Through the speculum it is observed to be anæmic or pale in color, and to have on its inner surface slight abrasions. The vagina contains some bloody discharge, which is not fetid. Ordered to have daily a drachm of liquid extract of ergot. After a fortnight, there being no diminution of the bulk of the uterus, and irregular hæmorhagic losses persisting, the cervix was dilated by tangle-tent. On the introduction of the tent, hæmorrhage began suddenly, and proceeded to an alarming extent, two pints being the quantity estimated as lost within fifteen minutes. Mr. Garstang injected through a hollow probe a drachm of tincture of perchloride of iron diluted with an equal quantity of water, with no result. A small fiddle-shaped india-rubber bag was now introduced within the cervix. It stopped the hæmorrhage. At 11 A.M., about thirteen hours after the hæmorrhage, the bag was a second time expelled. No recurrence of hæmorrhage. At 3 P.M. she was placed under the influence of ether, and the hand introduced into the vagina, two fingers with some difficulty into the uterus. On the posterior wall of the uterus was felt a projecting, moderately hard, wartlike mass of irregular form, and of extent equal to nearly two inches square. At first it was supposed to be a malignant outgrowth, but as a line was found at which it could be detached, it was recognized as placental. Some difficulty was experienced in removing it by a sawing motion of the nails of the fingers in the uterus. About eight ounces of blood were lost during the operation ; but afterwards there was only a moderate

amount of blood-tinted discharge. The mass was placental. On its
fœtal surface were only small patches of chorion. It was about a
third of an inch thick, and dense in structure. The section was grayish-yellow, and bloody, it being almost certain that blood had continued to circulate in some of the sinuses, so maintaining the vitality
of the mass. From these sinuses, where utero-placental, the flooding took place. The use of ergot was continued. Nine days after
the operation the uterus measured nearly three inches and a half only.
The cervix felt not more than half as bulky as it was. Fourteen
days after the operation the uterus measured two inches and a half,
and the cervix was reduced to natural dimensions.

This woman was very ill; her case was recognized as probably dependent upon miscarriage, although the miscarriage was the enormous distance backwards of eight months. I see no reason to think
that, if this woman had not been properly treated, she could have
escaped death from continuance of discharge; for the placental mass
was alive, and had firm adhesion to the uterus; and when separation
would have taken place I do not know. I think it would not have
taken place, but have led to the woman's being drained of blood, and
dying. The case was supposed to be connected with a recent miscarriage, because there was no evidence of fibrous tumor nor of anything else that would account for the bleedings and the great size of
the uterus. Had this woman's uterus been enlarged by a fibrous
tumor so big as to make the cavity measure five inches, the tumor
would have been easily felt; but no tumor was felt. The uterus, instead of being enlarged as it would have been by a fibrous tumor, was
a flattened mass which could not be distinctly felt through the anterior wall of this woman's abdomen. I call your attention to the great
size of the uterus. There was no need of this size to include such a
small thing as the bit of placenta which we took away, and the removal of which was followed by the complete cure of the woman, and
the diminution of the uterus to its natural size. The case, then, is a
very remarkable illustration of the power of a persistently attached
piece of living placenta in maintaining the development of the organ,
or, in other words, preventing its involution. In this it contrasts
with the comparatively small size of the uterus in the next case, that
of missed abortion. The case is quite clear. The woman had decidual endometritis affecting a part of her placenta, and making it
adherent. The placental decidual endometritis was probably also
the cause of her abortion.

Before leaving the case, I call your attention to the circumstance of the great rapidity with which the uterus returned to its natural dimensions, after the offending bit of placenta was removed. Fourteen days after the removal of the placenta, the uterus and its cervix had both returned to the natural size, after eight months of persistent hypertrophy.

The injection of perchloride of iron by Mr. Garstang was used before I had become satisfied of the danger of this remedy; arising from its sometimes passing into the veins, causing clotting of blood and embolism. In some such cases death would have resulted, if the embolism had been survived, from sloughing of the parts tanned by the iron.

I now come to the subject proper of my lecture,—missed abortion. Before entering upon that I shall say a few words explanatory of rare conditions that occur in connection with this department of obstetrics. Protracted pregnancy is entirely denied by some eminent obstetricians; I believe, however, in its occasional occurrence. Protracted pregnancy is the condition of a woman who has passed two hundred and seventy-eight days—the interval between the last day of last menstruation and the expected confinement—and at least a fortnight more than this. There is, indeed, no very exact definition of the number of days at the end of which pregnancy becomes protracted. If, at this time, a woman's child dies in utero, there is not then protracted pregnancy; she is in a state of missed labor.

It is necessary to say something as to this point—namely, when a protracted pregnancy ends, or when pregnancy of any kind ends, and the condition of missed labor or missed abortion begins. You cannot say that a woman is pregnant, without misleading your hearers, if she has only a lithopædion in her abdomen; neither is a woman properly described as pregnant who is in the condition of missed labor or missed abortion. This subject is of great medico-legal importance, as I shall show you presently.

Let me first give you the particulars of a remarkable case of protracted pregnancy and missed labor, which occurred in my private practice, and which forms a good illustration of these morbid conditions. The lady was forty-one years of age when she became pregnant for the first time. The uterus was, from the earliest time after its ascent into the abdomen, anteverted or pendulous. It was not the common form of pendulous belly, which can be replaced by bandage and held up; it could not be replaced. This impossibility of replacement was also observed during her confinement, and there was no reason to believe that there were any adhesions of the uterus. Her pregnancy up to the end of the natural term was otherwise perfectly healthy. She had a slight degree of generally contracted pelvis. Before giving you the dates I may tell you that none, in the most careful ordinary life, could be more accurately ascertained or more reliable than those I now state. Her menses ended on December 12th. On December 15th her husband left home, and did not return for nearly two months. Her confinement was expected on September 17th. The motion of the child ceased on September 26th. On October 17th she shivered and became feverish without any indication of labor commencing. It was considered necessary to deliver her. The mouth of the womb was artificially dilated, and she was artificially delivered on the following day, October 18th. The child was enormous—a female, dead. The mother died on October 24th. This is a case in which you have, with almost scientific certainty, slight protraction of pregnancy and then the condition of missed labor. After a fœtus's death, under any circumstances, it is generally discharged within a fortnight. In this case more than a fortnight elapsed after the cessation of movements, and there were never any symptoms of labor.

In some respects missed miscarriage or missed abortion is even more important than missed labor; for, in a case of missed abortion, the history of the woman and her size may have led either to no suspicion of pregnancy having commenced, or to suspicion which may have been dissipated by the further history of the case. In a case of missed abortion or missed miscarriage the important element of suspicion as to the real condition may not have come into the mind either of the patient or her physician. Mistake is then extremely liable to occur. This is not so likely in missed labor; for in that condition the woman's size will almost certainly have made her aware that she is in an advanced state of pregnancy; and her friends will also know it. I told you that missed labor may be a subject of great medico-legal importance. The same is true, and even more so, of missed abortion or missed miscarriage. Take the case that I am going to read, where a woman passed a fœtus of about two months at the end of a pregnancy (if you so miscalled it) which lasted for five months. If, in such a case, the practitioner, without sufficient care, were to tell the husband that his wife had had a two-months child, you can easily understand that his natural rejoinder might be, "That cannot be my child, for I have been away from home five months!"

Such unfortunate misapprehensions have happened, and the occurrence shows the importance of counting the term of a woman's pregnancy, not up to the time when the fœtus is discharged, but back to the time when it died. If this is kept in mind, the practitioner, in the imaginary case that I have given, will not make the mistake of leading the husband to think that the fœtus just born could not have been begotten by him. It is sufficient to allude to this, the medicolegal importance of it is so plain.

Now, when a woman has a missed miscarriage or a missed abortion, what is the course of events? The fœtus dies; the symptoms of pregnancy are arrested; milk sometimes appears in the breasts; hæmorrhages from the uterus may occur, or they may not. If the liquor amnii is not discharged it is absorbed, and the contents of the uterus either macerate or become mummified. If the membranes remain entire, the process undergone by the uterine contents is that of mummification. It is only when germs are admitted, and generally after rupture of the bag of membranes, that putrefaction and maceration take place, and the more or less complete dissolution of the ovum. If the uterus has been felt, the remarkable observation may be made, that a woman going on apparently in pregnancy has the uterus steadily diminishing in size, instead of getting bigger; and at last, and almost invariably (not invariably), before the full term of pregnancy, counting from the commencement of it, would have been reached, the ovum is expelled. The expulsion is frequently unexpected. When it is expelled, you have a mass in a state of mummification, nearly dry, of a dirty brown color; and the fœtus and membranes are concealed, being rolled up in the placenta, which is too firm to be compressed and embraces the whole ovum. Such ova I have had sent to me more than once by practitioners, saying, truly, that the fœtus appeared rolled up neatly in the membranes and the placenta as in a parcel. That was exactly the case in this instance. In this preparation you will see that the placenta and membranes have been opened up to show the fœtus inside. In our case the edges of the placenta met over the fœtus, embracing it entirely, rolling it up in a parcel-like form. I will now read to you the case.

S. K., aged thirty-one, married eight years, has had four children (the last two years ago), no miscarriages. Had not menstruated for five months when a bloody discharge began. After this had continued for three weeks she became an out-patient under Dr. Godson. She was ordered ergot and strychnine, and the discharge ceased. But it soon recommenced, and she came into the hospital. Examination now discovered a dilated heart with a mitral regurgitant murmur. There was dulness above the pubes for an inch, but nothing abnormal could be felt. Digital examination per vaginam discovered the brim of the pelvis occupied by a moderately hard mass, with which the cervix, which is patulous, is connected by continuity. The uterine probe passes easily into the uterus three inches and a half. The uterus is mobile, not tender, and forms the mass occupying the pelvic brim. About six hours after this use of the probe, which was withdrawn untinted by blood, pains began. After about eight hours of pains a mass as big as an orange was expelled. Very little hæmorrhage accompanied and followed the birth of the mass. The patient rapidly recovered. The mass was found to consist of the entire ovum in a state of decomposition; except the liquor amnii, of which there was not a trace. The whole presented a dirty-brown color, somewhat like that of decolorized blood. The decidua and other membranes were rolled tightly around the fœtus, the edges of the placenta meeting over it. The fœtus was of the size of about two months' growth. On the fœtal surface the placenta was covered with rounded projecting masses of various sizes, as of a field-bean, or of a hazel-nut. They were beneath the chorion, and were formed of blood-clot in various stages of decolorization.

This is as perfect a case of missed abortion as you could desire to see. The length of detention, after the death of the fœtus, is five months; the woman then began to feel herself ill because she began to bleed. Observe, in this case, that the membranes remained entire; therefore there was no putrefaction. The whole ovum was in a state of decomposition. Here I cannot avoid pointing out a common mistake in obstetrical writing. Some of the best books on obstetrics divide all children and abortions into living or putrid. That is a very great mistake. Dead children, dead abortions, in various stages of decomposition, are quite common; but putrid fœtus or putrid abortion is quite a rarity. Your nose is a sufficient instrument of diagnosis. A decomposed fœtus is very seldom putrid, and it should not be so described. In our case there was no putridity, but there was the peculiar condition of decomposition which I have called " mummification."

In this case I call your attention to what is perhaps a very important element,—the disease of the heart. It is only recently that great care has begun to be paid to the bearings of disease of the heart upon pregnancy and parturition; I know of none paid to the bearings of disease of the heart upon abortion. It is a subject well worthy of attention and study. It would be quite easy to erect a theory of this woman's abortion founded upon disease of the heart. Disease of the heart induces miscarriage frequently. This is not a case of miscarriage; it is a case of missed abortion; therefore, the explanation of the dependence of the death of this child upon the disease of the heart (mitral regurgitation) is far from being made out. This is, as I have said, a subject which, like innumerable others, remains for you to investigate.

You will notice in this case that I introduced the probe, and those who were present will remember that I said at the time, "I do this without hesitation, because, if the woman is pregnant, I wish the pregnancy to end." Before you decide to introduce a probe into the uterus you should always consider the question of pregnancy. In this case it was considered, and the probe was deliberately introduced. You see also beautifully illustrated, in this case, the power of what is called uterine catheterism in inducing labor. A single introduction of a uterine probe within six hours set the machinery of uterine pains agoing efficiently.

II.

ON ABNORMAL PELVIS.

THE subject of this lecture is abnormal pelvis. An abnormal pelvis is not necessarily a deformed pelvis; it may be merely a small one. A deformed pelvis may be, as you see in this example, both small and deformed. The most frequent deformity occurs in pelves that are not otherwise small—that are large enough except in the seat of the deformity. In connection with this subject we have a very great piece of progress in obstetrics that is going on at the present moment. Within my days, the introduction of anæsthetics into midwifery was a very great improvement. A still greater improvement, because saving of life is of more importance than saving of pain, has been the applications made of the antiseptic theory, not chiefly in the treatment, but in the prevention of diseases. That is undoubtedly the greatest improvement in obstetrics in modern times, and it is an improvement that is still going on and increasing.

The subject that I am now to lecture on is a part of the great improvement that has been introduced in the treatment of abnormal pelvis. To show you in one sentence the striking character of this improvement, I may tell you that while, not very long ago, I visited an obstetric hospital which was not possessed of a callipers at all—had not such a thing, nowadays, in many of the best obstetric hospitals, every woman is measured to find out the conditions of her pelvis. I am not recommending you to measure every pregnant woman, yet these measurings have resulted in very considerable increase of our information; and although this universal application of measurement is not required, still it shows you the contrast with the condition that I have mentioned of a hospital that had not callipers at all. This great improvement has been introduced from Germany, and it is, in the main, an importation from Kiel. In order that you may understand it, I use the old division of mechanically difficult cases into three. You have firstly the slighter cases—and therefore the more frequent, and in that respect the more important cases—where the pelvis is spoken of as a pelvis whose conjugata vera varies between four inches and a little above three. These are the slighter cases. Now, in these cases the improvement that has been made is an improvement in our judgment of the conditions of the labor—an improved diagnosis, so that cases which are still extensively spoken of as cases of inertia (which is, no doubt, generally an erroneous explanation, far too widely applied), or simply spoken of as "forceps cases," are now more exactly and correctly defined. They are recognized chiefly by deviations from the ordinary progress of labor, or from the ordinary mechanism; and these deviations from the ordinary mechanism are in a very great measure distinctive, especially of cases of mere smallness of the pelvis, the pelvis being otherwise well formed; and of cases in which the deviation of mechanism is produced by antero-posterior contraction of the brim without the pelvis being otherwise small. This is not the place to speak further of this kind of diagnosis made during labor. I merely point it out to you because I wish you to see intelligently the interest attaching to preliminary investigations generally and in the cases that I am to bring before you at a further part of the lecture.

If we come now to graver cases—the second kind of mechanically difficult labors, where the pelvis varies from above three down to, in exceptional cases, nearly two and a half inches in the conjugate. In such cases the great improvement which is still going on is an improvement, not in diagnosis, but in our judgment of the method to be pursued in delivering. In such cases it has been common, indeed, it may be said to be prevalent, for students or practitioners to divide themselves into two classes, and one set to swear a belief in version as the proper mode of delivering women with deformed pelvis, while another set believe in the forceps as the proper mode. All such judgments are ill-founded. They are founded upon the measurement of the conjugate as the criterion; and it was and is taught extensively that according to certain minute measurements of the conjugate, so you should proceed to deliver a woman by podalic extraction after version, or by forceps, or by craniotomy. Such a method of judgment must be entirely given up. It is necessary nowadays, if you are to treat your patients properly, to come to each case unprejudiced, to study it as an individual case, in which there are a great many elements besides the mere measurement of the conjugate, some of them more important than any refinement of that measurement. Among these elements are the presence or absence of general contraction of the pelvis, the position and other relations of the head, the state of the membranes, and the state of the uterine retraction. Now I wish to point out very impressively this error which leads a man to treat a case on the assumption that all he has to do is to measure the conjugate.

A similar defect in judgment runs through the recent writings in favor of the increased frequency of the use of the forceps in what may be called ordinary labors. In the case of deformed pelvis it is the measurement of the conjugate that is held to be the criterion of practice—the better judgment founded on the consideration, not of one, but of all the important elements of the case, being omitted or lost. In the case of forceps, statistics, whose accuracy requires consideration, are held as showing success resulting from a certain frequency of their use, and practitioners are directed to look at that frequency as a criterion of good practice—the better judgment founded on a full and careful consideration of all the particulars of each case, or of each group of cases, being again also omitted.

Although it is out of place, I shall here make one remark on using statistics in judging of the forceps practice referred to. The forceps cases of a forceps enthusiast are unfairly set against those of one who rarely uses the instrument. I advise you to trust to nature as far as you wisely can; to be loath to take a case into your own comparatively ignorant and unskilful hands; and to judge that the success which the forceps practitioner seems to have, as against leaving cases to nature, is a fallacious appearance of success, if it be true that nature is on the whole better than forceps. For, if a forceps practitioner delivered all his cases artificially, his so-called success would be still greater, which is absurd. Practices in which the forceps is often used should be compared with practices in which the instrument is rarely used. We require more diagnostic refinement of the causes and conditions of difficult labors; and it is a part of this diagnostic progress that I am trying to teach you to-day. This improvement will diminish the number of cases going by the name of the treatment, as forceps, and describe them less nosologically and more pathologically. No doubt it will diminish also the number of cases vaguely called inertia, or declared to be from an undiscoverable cause.

The third class of cases—the gravest cases—cases which run from two and a half inches downwards to less, have also undergone very great improvement, the improvement being in the kind of instrumental treatment, the means of carrying out the design of the practitioner; not as in the former class, deciding what is to be done, but the method of doing it. Upon this third class I shall say nothing more meantime.

Now I come back to the first set of cases, which are far the most important—the slightest class of mechanically difficult labors. The astonishing result has been clinically arrived at, that in Germany there is a mass of from 12 to 15 per cent. of such cases. I am quite sure that there will be found much fewer in this country. That is a judgment, not a statement founded upon exact information, because I know no hospital or practice in this country where there has been systematic measurement of every case and observation of the mechanism of early labor, with a view to decide such a question; but it is founded upon this, which is almost positive proof, that in this country malpresentations, cord presentations, face presentations are rarer than in Germany. I should be very much astonished, therefore, if a careful clinical inquiry resulted in showing that in this country there were so many as from 12 to 15 per cent. of pelves abnormal, as has been found by thoroughly competent authorities in Germany.

In these slightest cases, pelvimetry is most difficult. The pelvimetry in these cases consists in very simple measurements, which, however, you require to learn to make. A practitioner is very awkward in making such measurements at first, and he requires to have a good callipers or other good external pelvimeter. He requires experience, still more, for internal pelvimetry.

How do you proceed in these cases? The patient is undressed, and placed on a suitable bed for examination. The object is to find out as nearly as you can the length of the conjugata vera, and to find out the general size of the pelvis. In all cases these are the two chief things; but in cases of higher deformity you go farther, and measure such things as the distances of the posterior superior spines of the ilia, and make a variety of further observations which I do not enter upon now.

The first measurement is of the external conjugate ("C. ext."), frequently known as D.B., the diameter of Baudelocque. Now the external conjugate is measured from what you judge to be the first spine of the sacrum, or from a hollow that is generally found below the last lumbar spine, to the mons veneris in front of the symphysis pubis. In a healthy woman that measurement is from seven and a half to eight inches; I shall put it down at seven and a half. There are sources of variation which will easily suggest themselves to you, such as the different amount of fat in different women. Now for the judgment you form from this. You take off two and a half for the thickness of the sacrum ; you take off fully an inch for the thickness of the pubes and the soft parts—that is, you subtract quite three inches and a half from seven and a half. If you had nothing else to rely upon, and you found the measurement seven and a half, you would say a four-inch pelvis—a healthy pelvis so far—four inches. But you will find in practice that this is not a very reliable measurement ; therefore, you take other measurements by which to correct this. It so happens that in the most interesting case I have to mention to-day the measurement proved correct, or as nearly correct as was to be expected. In this poor woman, whose pelvis I have in my hand, the external conjugate was five and a half ; take off fully three and a half, and you have left two inches, or somewhat less.

The next dimension you take is the measurement of the spines, as it is called. This measurement is from the external margins of the anterior superior spinous processes of the ilia, and it is known by the marks I show you here, "Sp. il." In healthy women this measurement varies greatly, and it is about ten inches. Then you take another measurement between the most distant parts of the crests of the ilia, and this is known in books as "Cr. il.," and in healthy women generally measures eleven inches, or fully an inch more than the former. These two measurements afford valuable evidence ; they are easily taken, and you will find their value excellently illustrated in the cases I have to go over immediately. If these measurements are both small, then you have reason to suspect that the brim of the woman's pelvis is small. If the measurement (as in this case) of the crests is smaller than the measurement of the spines, or equal, then you have reason to believe that the pelvis is contracted or flattened in its antero-posterior diameter.

The next measurement is the most difficult ; it is also the most important. In the graver cases no other measurement is absolutely required—that is, the measurement of the conjugata diagonalis, which is known in books as " C. d."—generally in a well-made pelvis four inches and a half. But in a full-sized pelvis it is often not to be measured during life ; to do so would give the woman too much pain ; you would have to force the fingers too far in order to succeed. You will see how easily it is measured in some of the cases of contraction that I shall presently describe. This measurement is made by pushing one or two fingers per vaginam so as to touch the promontory with the point of the index finger if one is used, or of the middle finger if two are used (the index finger being not long enough). With the nail of the index of your other hand you mark off where the inferior border of the symphysis cuts the radial side of the introduced index finger, and then you have a pretty accurate measurement of the conjugata diagonalis by telling off the distance between the point of the index finger if that alone was used, or between the point of the middle finger and the mark you have made with the nail of your other index upon the radial border of the hand. This gives you the conjugata diagonalis. Now from this you argue as to what you wish to ascertain; namely, the conjugata vera ("C. v."). The conjugata diagonalis being ascertained, from this take half an inch, and you get the conjugata vera which you seek. There are a good many niceties about this measurement, but you get as your result in a healthy pelvis four inches from this plan, just as you get it from the diameter of Baudelocque.

These measurements, in the slighter class of cases, are important, but they have to be supplemented by measurements during labor, and by observations of the mechanism of early delivery.

Now I come to the cases. We have had recently in "Martha" four cases, not of the first or slightest class, but of the second and third.

The first case is one of which this well-known museum preparation may be held to be a representation, for in the patient, whose case I have now to read, the condition was exactly similar. The case is one of osteo-sarcoma of the sacrum; the pelvis being neither small nor deformed, in the ordinary sense of those words; but for obstetric purposes extremely deformed.

E. P., aged twenty-seven, married for seven years, has had four children, all born at full time; complains of almost constant pain in the lower part of the back, greater on the left than on the right side. This pain has been present since her last confinement, seventeen months before admission into the hospital. About the seventh month of her third pregnancy she first felt this pain—about three years ago. The child was delivered by craniotomy. The pain, which had been less or altogether gone, returned about the seventh month of her last or fourth pregnancy. This child was also delivered by craniotomy. Besides the pain she has leucorrhœa and frequent micturition. She has not had a monthly illness for two months and thinks she is pregnant. She is on the whole a well-made woman. A large solid tumor occupies the posterior parts of the pelvic cavity so as to reduce the available conjugate to one inch and a half or thereabouts. There is a rounded, flattened, and slightly projecting swelling of the base of the sacrum externally, and more on the left than on the right side. The uterus is elevated above the brim of the pelvis, and is three inches in the length of its cavity. She was found to be not pregnant, and was dismissed.

You will observe this case was not measured by callipers, because measurement by callipers could afford us no information—the woman had no deformity to be detected in that way ; and besides, the external tumor would render any measurement by callipers useless. The fingers here made the measurement : they measured the available conjugata vera actually and at once, and they found it one inch and a half at the time of her coming into the hospital. Here the measurement of the conjugata diagonalis was not attempted, not required, and it could scarcely have been done. This woman's disease began before the third pregnancy, in which she was delivered by craniotomy, after having had her former children easily enough. The disease was gradually increasing ; and now, if she were falling in the family way again, abortion should be induced to save her from the dangers of delivery by Cæsarian section. She could not be delivered, if she went on to near full time, in any other way. In this woman, then, had we found pregnancy to exist, we should not have hesitated to destroy the pregnancy in order to save her from the dangers attendant upon delivery of a child at or near term.

Cases of osteo-malacia are very uncommon in this country. There is a case at present in one of the medical wards. A woman may be seized with this disease after she has had some children quite easily, and may offer you a history like the history of this woman, of gradually advancing deformity of the pelvis. But in the case of osteomalacia you would have very different conditions. The whole skeleton is modified, and the woman is gradually sinking in stature as well as having her pelvis diminished in its conjugate diameter. In fact, the cases have no analogy to one another except in the circumstance that you have the deformity of the pelvis gradually increasing from one pregnancy to another, and requiring, as the deformity advances, different kinds of delivery if the woman is allowed to go on to full time.

The next case is one of a commoner kind—a case of generally contracted rickety pelvis. This woman, aged twenty-seven, was brought into Martha Ward in labor on June 24th last. She has been deformed since childhood, and is of low stature, measuring four feet two inches. She was married on September 24th, and has had no catamenial discharge since then. Pains began on the 22d; they were never severe. The cord became prolapsed on the morning of the day of admission—it is pulseless. The diameter of Baudelocque was found to measure five inches and a half; the crests measured eight inches and a quarter, while the spines measured more—eight inches and a half. The uterus has a natural feeling, projects extraordinarily, and has a left lateral obliquity. Through the hypogastrium the child's head can be felt, movable. The limit of the uterus and cervix not distinctly felt, from the pains being slight—it is about half an inch below the level of the navel. The external parts are swollen and congested. The external os uteri is dilated to the size of a florin. The head presents in the first position. Two fingers can with difficulty be squeezed into the conjugate, which is almost an inch and a half, and there is no considerable increase of any antero-posterior diameter of the brim at any part. Some pelvic brims have dilatations at one or both sides of the promontory; in this case there was no increase. Cæsarian section was performed, and proved fatal from septic peritonitis of slight extent and degree upon the third day. In this case the callipers were used, and they alone indicated very accurately the kind of deformity and the degree. But the fingers gave an additional measurement by being jammed into the actual and available conjugata vera, so as to measure directly the size of the conjugata, just as in the former case. The pelvis in this case was made out without any difficulty to be a pelvis which was generally contracted or small, highly deformed, with a conjugate of an inch and a half, and its deformity was rickety, the brim having a reniform or kidney shape. I have not entered in this case upon the woman's medical history, which of itself showed that she had a pelvis almost certainly rickety, and involving great difficulty and danger should she come to be confined at or near the full time.

The next case I have to mention is one of the commoner kind; it is also a case of generally contracted rickety pelvis. This young woman was aged twenty-two, healthy-looking, four feet four inches in height; had her last monthly period in the beginning of April, six months ago; had previously been always regular. The legs are curved, nearly symmetrically, the convexity looking outwards to

either side, the greatest curvature being at the junction of the middle
and lower thirds. The abdomen presents the characters of a pregnancy advanced beyond the sixth month. The posterior superior
spines of the ilia are not easily or well made out—two inches apart.
The diameter of Baudelocque is six inches; spine eight inches and a
half, the left being an inch and a half higher than the right; the
crests eight inches and a quarter; the diagonal conjugate is three
inches; the sacrum is acutely bent in a posterior angular curvature
below its middle. The spine has a slight right lateral curvature in
the dorsal region, compensated by one in the lumbar region to the left.
The induction of premature labor is recommended as soon as the
child is viable, the conjugata vera being judged to be little more than
two inches and a half.

You will observe the words I use in regard to this case: that the
conjugata vera is "judged" to be so-and-so. In this case you cannot, before labor, actually measure it—you cannot measure it, as in
the two former cases, by jamming the fingers into it. In all cases
that can be done just after the child is born, and should be done. In
the great deformities, such as those of the two women I have previously described, it can be done before labor, but in a case like this
it cannot be done. Therefore I have here a judgment as to the
measure of the true conjugate; we do not actually measure it.

I have still another case of equal interest, but I shall not read it to
you. I shall merely mention it. It is like the last, but still slighter
in its dangerous character. It is the case of a woman who had had
eleven children, and of these children she bore only two spontaneously
—the first two. Of these two the second alone was born alive, and
survives. Now I mention these few particulars of this case to point
out to you an observation of great interest—the contrast between successive labors in a slightly deformed pelvis and in a healthy pelvis.
Everybody knows that, in an ordinary practice, tedious and difficult
cases are expected among the primiparæ; and it is quite true. The
observation is correct. In the cases of primiparæ you are not astonished at having a long, expectant sederunt. Subsequent labors are
undoubtedly more and more easy, mechanically speaking, till at last
they very frequently become far too easy for the woman's safety. But
in the case of the first degree of deformity of the pelvis you have, as
this case illustrates, the opposite course. It is the first labors that are
easiest. In the first labor the woman's power, and especially the labor
including the uterine power, is the greatest, and in a woman's first

labor she may succeed in forcing the child at the full time into the world, while in subsequent labors she utterly fails from weakness or inadequacy of the powers of labor. In a woman with a slightly deformed pelvis you expect subsequent labors to be the more difficult, apart from any increasing deformity, and simply from the powers of labor being less as pregnancies increase in number—a well-known fact.

I go back to repeat what I said when I was speaking of the second class of deformed pelvis, that the measurement of the pelvis, and especially the measurement of the conjugate, even if accurately made, is not the criterion of the mode of delivery to be adopted at the full time, or if premature labor is induced. In the same woman, conditions may vary in different labors; and, in different cases of the same dimension, conditions may vary, so that at one time perforation may be the right operation, and at another time turning may be the right operation; and I may state to you that turning, or rather delivery by podalic extraction after turning, is not to be resorted to unless you have a rational prospect of getting a living child. If your delivery by turning ends in the birth of a dead child it is, to a considerable extent, a failure; it would have been better to perforate—safer for the woman. You may not justly condemn your practice retrospectively. Nevertheless, it is a fact that you would not choose to turn a dead child; and if you turn a living one, and do not extract it alive, your operation is partly a failure; perforation would have been better.

You may use the forceps, and you can easily understand that not only may the forceps be used in one instance in the same woman, where in another instance turning is the right operation; but you may be pretty sure that as the forceps is the operation most used in the slightest cases, so it will be the most frequent operation; you will more frequently have recourse to the forceps than to podalic extraction after version; but that frequency is nothing at all in favor of the forceps as an operation in jealous rivalry with version. There is no just occasion for any rivalry. Every case must be judged of on its own merits, the whole particulars being taken into consideration.

Now, finally, suppose you have had a case of this kind. The future treatment of the woman is easier, because in future pregnancies you have the history of the labor in the former pregnancies to aid you. And every woman who has a deformed pelvis should have kept for her a careful record of the history of her various deliveries, so that the practitioner may have the instruction derivable from former deliveries.

Every woman whom you deliver who has a pelvis that is at all suspected of contraction, should have five different measurements of her brim, for the purpose of guiding the treatment in subsequent confinements.

Firstly, you have the measurement of the conjugata vera founded upon the measurement of the diameter of Baudelocque—the external measurement; and that you can get at any time. Secondly, you have the measurement of the conjugata vera founded upon the measurement of the conjugata diagonalis; and that measurement you can frequently get at any time, whether the woman is pregnant or not. The third measurement is a measurement that we can only get when the woman is not in a state of advanced pregnancy; it is a measurement which is easily made in a thin woman—a woman who has not much fat in the anterior abdominal wall, nor any kind of abdominal distension. You can make out in such a woman through the anterior abdominal flap the promontory of the sacrum and symphysis pubis, and measure the intervening distance. Then you have a fourth measurement, which generally can be made, and is made, only during delivery or immediately after it. I told you that in a slightly contracted pelvis you cannot actually measure the conjugata vera before delivery as you can measure it in an extremely contracted one by jamming the fingers into it; but immediately after delivery it is your duty to do that, and you do it by introducing your whole hand into the pelvis. Every practitioner knows the breadth of his hand at different parts, and he finds out the number of fingers he can pass into the conjugate, or the degree to which his whole flat hand will go into the conjugate. He can measure actually at that time the size of the conjugata vera. That is a fourth measurement that every woman should have made upon her during or after her labor if her pelvis is suspected. There is a fifth which is also very valuable. Of course, in a case of delivery of this kind, you watch the passage of the child's head, noticing the diameter which comes through the contracted part; and, as soon as the child is born, you take your callipers and measure this part, generally near the bi-temporal diameter, and you measure it, pressing your callipers pretty firmly, as probably the pelvis pressed pretty firmly as the child's head came through. This gives you the size of the body that came through.

III.

CHRONIC CATARRH OF THE CERVIX UTERI.

THE case which forms the subject of this lecture is one of chronic catarrh of the neck of the womb, a disease which has, for many years, been popularly known in the profession and to the public, as "ulceration." This term conveys a very erroneous idea of the formidable character of this disease, so that it has given to patients an immense amount of unnecessary and unjustifiable alarm. When a woman has a genuine ulcer of the womb, such as would be so designated by a surgeon, destroying tissue deeply, you have ground for alarm, for most of these cases are malignant in character.

The disease is now generally called by the name I have given it.

First, it is chronic. You are all familiar with acute forms of catarrh, such as the common cold in the head, which for a few days causes so much fever, pain, and annoyance, and then disappears. A woman is liable to similar acute catarrh of the cervix uteri; but that is not the disease of which we are speaking. Our disease is chronic, for it is of long duration, sometimes being so even when diligently treated. It may last for years or a great part of a lifetime, during which a woman may have borne several children. In the case now in Martha Ward we judge from the history that it has lasted, at least, thirty-two weeks.

Second, it is a catarrh, presenting all the usual appearances of this diseased condition. The mucous membrane is swollen, red, is easily made to bleed, and secretes a muco-purulent fluid or simple pus. In the part of the cervix that can be seen, the mucous membrane has often a punctuated appearance, which is called granular. This arises from epithelial denudation or so-called ulceration, laying bare or making visible the papillæ, which are specially injected, and whose vessels are easily ruptured, and bleed. Our patient complains greatly of losing blood; so much, indeed, as to be called by her (not by us) flooding.

Third, it is an affection of the neck of the womb. This part, you must always remember, is physiologically and pathologically, as well as anatomically, quite distinct from the real womb, or body of the womb. The latter is the organ of menstrual excretion and of pregnancy. A neck of a bottle is much less a distinct part from the bottle proper than is the neck of the womb from its body. The cervix uteri is a large open gland, and very liable to catarrhal inflammation.

This, then, is the disease, chronic catarrh of the neck of the womb.

This disease is of considerable importance on account of its frequency, not on account of its nature. It is in every respect an important disease, yet it is not to be classed with fevers, degenerations, with rheumatism, or gout. If a classification of diseases were made, according to their gravity, I dare say this disease would not be placed higher than the third rank. Many women—but far from all—who suffer from it, pay no attention to it, and can scarcely be said to be patients in any ordinary sense. In some women it is important from the alarm it causes; in our patient in Martha Ward, it was supposed to be a malignant rodent ulcer. In all it deserves attention, and demands treatment at your hands.

It is generally said to be the commonest disease peculiar to women; but I am not sure of this. I think it has rivals in chronic ovaritis, and in chronic inflammation of the uterus and ovaries. Yet there is no doubt its commonness justly gives it great interest and prominence.

I cannot pass on without saying a few words on the historical position of this disease, "ulceration of the womb." This history is an illustration, and in some respects not a creditable illustration, of the medical philosophy of this century. It shows that the period of medical enthusiasms, not yet passed, has characters, besides those of weakness, allying it with passing religious enthusiasms. Ulceration was raised into the position of a gynæcological system, and all the diseases of women were managed accordingly. I can well remember —indeed all except students cannot fail to do so—how, over the whole world, gynæcological practitioners were busy with speculum and caustic, and thought they had in these tools a panacea for the diseases of women.

Luckily for you, great medical systems are unknown now. Had you been students a generation or two ago, you would have been taught, as your paramount acquirement, a system—of Boerhaave, or of Cullen, or of Broussais; and you would have been carefully indoctrinated, it being held that you could not practice safely without the guidance of a system, and that in all your dealings with your patients you should keep the system before you as your guiding star. Just so was it with the little ulceration system in gynæcology. We must stamp out these premature systems in medicine, and in gynæcology too.

The reintroduction of the speculum in the early part of this century, by Récamier, showed, as a striking and frequent phenomenon in women, a redness around the os uteri, which was called an ulcer. This discovery is the real commencement of modern gynæcology. It ripened into the system I have spoken of. This system is happily obsolete, and we can now calmly describe this important disease. Before leaving this subject let me give you a picture, almost in the words of one of the most eminent European gynæcologists, of the exaggerated views entertained, not above twenty years ago; and I may tell you this picture was regarded as no exaggeration by many, if not most, of the great gynæcologists of this country.

He gives a description of the fearful results of uterine catarrh and so-called ulceration, and blames the neglect of practitioners to examine—a blame which he carefully extends to the management of those cases wherein all bad symptoms having disappeared without local treatment, he declares the cure to be only deceitful and a source of dangerous confidence. He also expresses his conviction, that in at least eight out of every ten cases of hysteria the various nervous lesions depend on some kind of uterine catarrh, and impresses on his medical brethren that in no case of nervous disease in the female does he commence treatment until he has himself made a careful vaginal examination. A few of the nervous lesions he enumerates, including nervous headaches, hysterical affections, palpitation of the heart, neuralgiæ of all kinds, the most various spasms, hyperæsthesias, anæsthesias, paralyses of the lower extremities, etc.!

Chronic catarrh is very indistinctly referable to certain causes. Among them may be enumerated childless marriage, abortion, or full-time delivery, or cold, or gonorrhœa, or to suppression of the menses, as in the case immediately before us.

The patient complains of pain in the back, or, to be more exact, about the base of the sacrum. This is a common seat of cervical uterine pain, and is well illustrated in the pain experienced by women in labor during the dilatation of the os uteri. Pain down the thighs, feeling of weight about the rectum or lower part of the belly, are common; and there are many other ill-defined symptoms referable either to the disease or to the constitutional disorder which sometimes it induces.

What chiefly attracts the woman's attention in most though not all cases, is an extraordinary discharge. All such discharges, when not bloody, are familiarly termed "whites" by women; but, if there is any occasion to be exact, you cannot rest satisfied with such a mere name. You must see the discharge before it has dried on a cloth, or see it *in situ*. Often, and even in severe cases, there is little discharge to show. In our present case, although the disease was extensive, there was little discharge; it was only to be well seen by exposing the diseased part, and observing its thick, yellow, viscid character. You cannot judge of these discharges when dried on a diaper, for then they are all very nearly alike, appearing as dirty, grayish-yellow stains. A discharge in cervical catarrh may vary from the healthy crystalline viscid mucus of the part through opalescence to yellowness or greenness. The worst kind is not viscid, but a thin yellow pus.

A milky-white discharge is scarcely to be called morbid. It is the vaginal mucus in excess, and occurs in very many weakly women after a long walk, or even without apparent cause. A glairy albuminous crystalline, or slightly opaline, discharge is also scarcely to be called morbid. It comes from the cervix. But a yellow or purulent discharge surely indicates disease.

This discharge is to be traced to its source, and this is done by using a speculum, which shows part of the catarrhal surface with the discharge flowing. The discharge may be wiped off by a mop to disclose the disease better, and often the mop sets agoing an oozing of blood. The duck-bill speculum is the best, but it is not generally used in private practice, because it requires special adjustment of the patient and of the light, and the aid of an assistant. Besides some exposure of the patient's person is scarcely to be avoided. After it, the best speculum is the mirror-glass speculum, which I show you. These specula are made of various sizes, and you use the largest that you can introduce without difficulty.

The speculum only shows you a part of the disease, the part that used to be called the ulcer. It is now known that the disease often, indeed generally, affects the whole cervical surface, and in some cases, as in the one now in Martha Ward, the neck of the womb is so softened, its muscular coat so relaxed or paralyzed, that you can, by a probe or spatula, open up the external os, and look into the cavity of the cervix. This makes the disease appear very extensive. In most cases the opening up of the cervix is impracticable.

I will now read you some details of the case. M. D., æt. forty-six, married twenty-two years, four children—last seven years ago—was admitted on January 8th.

She says her catamenia commenced at sixteen years, with intervals of three weeks between each period, and continued fairly regular up to the birth of her last child, seven years ago, whence she dates her present illness. Her catamenia then became more profuse, and recurred with intervals of fourteen days. Thirty-two weeks ago the catamenia stopped for ten weeks, and a yellow discharge came on.

During the last three weeks she has had severe sacral pain. She has also pain in the hypogastric region, and shooting pains down the thighs. She has also had a flooding, which lasted twenty-four hours, and she continued losing slightly for fourteen days after.

Per hypogastrium, nothing unnatural is found. *Per vaginam*—Cervix uteri is in normal situation, considerably enlarged by expansion, so that two fingers can easily be introduced; quite soft, and partially denuded of epithelium, and secreting a viscid yellow muco-pus.

We will now proceed to the treatment of the disease, and this varies according to its severity.

In many slight cases a lotion may be used, partly to keep the vagina clean, and also to wash the accessible part of the diseased surface. The lotion may be applied by the patient herself with an ordinary Higginson's syringe, which throws the lotion against the cervix.

The lotion is used daily while the monthly period is absent, and often nothing more is required in the way of treatment.

It is a great mistake to use strong astringent lotions of alum or of decoction of oak-bark, for these have only temporary and only apparent good effects. They are injurious by the irritation of the vagina which they produce. A soothing, healing, cleansing application is what you want. Eight ounces of tepid water, holding in solution half a drachm of sugar of lead, is a good lotion; or the same water with half a drachm of alum, and also of sulphate of zinc.

The ordinary treatment is the cauterization, by nitrate of silver, of the diseased surfaces. The stick is to be passed into the cervix and turned around. This may be repeated every third or fourth day for several times. It is not the most successful treatment. Many cases do not yield to it; and frequently the practitioner perseveres with its use, not only long after it has ceased to be useful, but when it has become positively injurious. I have known this kind of treatment continued for years. Long before such a period has elapsed, indeed after several—say about ten—applications at most, in ordinary circumstances, the practitioner should have the case cured, or give it up as not amenable to the method.

In the severer cases, such as that which is the subject of this lecture, the best treatment I know of is by zinc-alum. This caustic has the advantage of requiring generally only one application. It was introduced to my notice by the late Dr. Skoldberg, of Stockholm. Sticks of zinc-alum, from one to one and a half inches long, are made by fusing together equal parts of sulphate of zinc and sulphate of alum, and running the mixture into moulds of the size of a No. 6 or 7 bougie. The cervix is exposed, and a sound is passed to find if the passage is clear, and to show its direction. Then the stick of zinc-alum is introduced and left in the cervix. A plug of cotton or lint is placed in the upper part of the vagina to keep the stick from coming out, and to receive the dissolving caustic. After three hours the plug is removed, and the vagina well washed with tepid water. The caustic produces a yellowish-white slough, which, after several days, comes off, leaving in successful cases a surface which secretes healthy cervical mucus, and soon assumes its healthy appearance. This has been the history of the case now in Martha Ward. The cervix is contracted, the catarrhal condition is nearly healed, and the secretion is healthy, all within a fortnight, from the use of the remedy once.

Zinc-alum is stronger as a caustic than nitrate of silver; as usually applied, it produces a very thin scale of slough, whereas the slough of zinc-alum is as thick as a sixpence.

In the severest cases, when you have hypertrophy and sometimes a nodular condition of the cervix, stronger caustics are of most use. Caustic potash, duly applied so as to produce a slough in the thicker hypertrophied lip, is the best remedy. Sometimes the actual cautery proves very efficacious. Cases of this kind are not cases of simple catarrh, but are complicated by peculiar pathological changes. They are sometimes now called erosion, or erosion with hypertrophy.

Finally, there are many very slight cases in which you have no morbid secretion, but merely a little red patch, often called an abrasion, on one lip or around the os. In such you had better not interfere. You unjustifiably alarm your patient, and you do her no good. Indeed, it is almost certain that local treatment will make matters worse. Bathing and other constitutional remedies may be resorted to. Such redness around the os uteri I have seen in an adult fœtus which had never breathed. Analogous conditions are frequent in the throat, and frequently subjected to prolonged treatment in vain. I have said that chronic catarrh is important ; and have, in concluding, to add that it is advisable you should not go on indefinitely treating it. If, after two or three trials, which may each extend over several weeks, you fail to effect a cure, you had much better give up further meddling in the matter. You do no good to the disease or to the patient ; you may, indeed, by frequent and prolonged irritation, produce a tendency to cancer.

IV.

ON OVARITIS.

THE diseases to which I am to devote most of this lecture are very difficult of precise investigation. They are seldom fatal, and consequently seldom illustrated in the post-mortem theatre. For these reasons progress in this department of gynæcology has been very slow.

Ovaritis, like several of the diseases that I have been lecturing on in this room, occurs as a complication of pyæmia; and such ovaritis I do not consider now at all. A case of this sort occurred in " Martha," and I just mention it to give you an example of kinds of ovaritis that I am not considering at present. A woman was delivered with great difficulty on account of placenta prævia, complicated with slight contraction of the pelvic brim. After delivery she suffered from putrid intoxication or sapræmia. She was in this condition brought to the hospital. The putrefaction was arrested by intra-uterine lotion. She then showed symptoms of pyæmia, and died. Two pelvic collections of pus were found—one (perimetric) in a cavity bounded by the left ovary, the uterus, and above by a piece of omentum: this contained half an ounce of pus; the surface of the ovary was ulcerated where it formed the wall of the pus sac; the ovary itself was enlarged, and corpora lutea in it contained pus. The other collection of pus was in the cellular tissue (parametric), to the right of and a little behind the cervix uteri, and contained the same quantity of pus. The walls of the uterus were flabby. The decidua serotina was easily seen. The right ovary was swollen, renitent, as big as a walnut, and when cut into was found to have its healthy tissue everywhere utterly destroyed and converted into a yellow, purulent, almost diffluent mass. There was no lymph in Douglas's space. Bladder and uterus normal; no general peritonitis. Of such ovaritis with suppuration examples are not rare, because puerperal pyæmia is not rare.

Leaving that subject I come to say a few words on what may be regarded as prefatory to ovaritis—ovarian irritation, often called ovarian neuralgia; a very common affection. It is characterized by absence of every sign of disease, and of every regular symptom, except pain in the region of one or other ovary; curiously, more frequently in the left than in the right ovary. You know that when a disease is characterized by pain, and nothing else, it is called an irritation or a neuralgia; and so it is in the case of the ovary. Although that is the name given to it, you must not suppose that that is the final or true pathology of the disease. I am very doubtful of that. In accordance with the nature of this disease, characterized, as I have said, by a pain in one or other groin over the ovary, it is treated by anti-neuralgic medicines, and a combination of zinc and quinine has had great reputation in cases of this kind.

I go a step further, and now mention to you some cases which are not inflammatory, and yet which are certainly more than merely neuralgic. Of this kind of disease many examples occur. Either one or both ovaries may be felt, and they are slightly enlarged—they are tender, and you may well ask, "Why, then, do you not call that inflammation?" There are the following reasons for this: That in many cases, as in one which I am about to read to you, the women are in perfect health—not in all cases, for many practitioners would ascribe the nervous, hysterical condition of some women to this disease of the ovaries; but it occurs frequently in women who are in blooming health, who have nothing to complain of except tenderness of the ovaries, not pain in them. Further, cases of this kind are not in any marked manner benefited by antiphlogistic treatment. They generally, indeed, resist all treatment. Here is a case: A. H., aged twenty-four years, married a year and a half; never pregnant; catamenia regular. She complains of painful menstruation. On examination the left ovary is easily felt and somewhat swollen and tender. The uterus is natural, except extreme sensitiveness of the mucous membrane of its body. The cervix permits easily the passage of only a No. 7 bougie. After some partially successful treatment of the dysmenorrhœa, she left the hospital, but soon returned, saying she was not cured. Now she privately made known that what she wished cured was not so much her painful menstruation as pain in sexual connection—a pain which delicacy had prevented her from earlier mentioning. With this in view, she was re-examined, and now both ovaries, somewhat prolapsed, swollen, and tender, yet freely mobile, were easily felt. Pressure on either of them produced pain, which she recognized as that of her dyspareunia. She is now under treatment. Counter-irritants externally, and small doses of corrosive sublimate internally, are being used. I can only say I hope she will be cured.

Now I come to cases about which there can be no doubt, where you have every local indication of inflammation that you can get under the circumstances, and corresponding constitutional disturbance. Here there is a very great difficulty, namely, in arranging the cases into classes according as they are acute, subacute, and chronic. Many cases are easily recognized as chronic and subacute; and among such some pathologists might place the case of dyspareunia that I have just read. But the great majority of cases it is impossible to place in any distinct category. At the other extreme, however, you have cases which end in abscess. These cases are evidently acute inflammations. Such acute cases ending in abscess, and not of the pyæmic kind that I have mentioned at the beginning of this lecture, are not uncommon, and the abscess is perioophoric, or in that part of Douglas's space in which the ovary is lying. Acute inflammation of an ovary after abortion, delivery at full time, or in the unimpregnated female, not rarely ends in pelvic abscess, which is a perioophoric abscess. No doubt, apart from pyæmic suppuration, cases of abscess or suppuration within the ovary do occur, especially small abscesses; but, clinically speaking, I know nothing of such cases. I have seen them in the post-mortem theatre. I am inclined to throw considerable doubt upon the histories of cases described as of large abscesses within the ovary, containing half a pint or pints of pus; and it would require careful dissection of such a case, careful consideration of all its history and characters, before it could be held as proved that such a large collection is an ovarian abscess, and not a suppurating cyst. I do not deny the possibility of a small abscess forming in the ovary; but I do not think it is proved that large abscesses, such as are described, ever form in the ovary. I have never seen a case nor a dissection which satisfied me on that point.

I have said that the progress of this department of gynæcology is very slow; and we owe progress in it to two causes—the recent enthusiasm in gynæcology which is prevailing all over the world; but more especially to increased exactness, and more extensive application of the bimanual method of examination. You know very well, and it is admirably illustrated in many parts of medicine, that although a thing is before your eyes you may not see it unless you look specially for it. Nothing is more painfully true than this; and

in no department of science is it better exemplified than in medicine.
When you carefully look for disease of the ovaries, and the careful
seeking is done mainly by bimanual examination, you will find that
there are very many cases of minor diseases of the ovaries. Biman-
nal examination may be made in various ways. It is to be done in every
case in which you pretend to make a thorough examination of the
internal genital organs. Before proceeding to it, the bowels must be
emptied—that is to say, it is, as a preparatory measure, well to give
a dose of castor oil. If you do not, you may find the visit to your
patient lost in consequence of the distension of the bowels with fæces.
You cannot make a fine examination of the ovaries while a quantity
of fæces is stuffing the pelvis. The bowels being emptied, your patient is placed on her back at the right side, or at the obstetric side,
of the bed; her right thigh is raised, the right foot being placed ad-
jacent to the left knee. With the forefinger of your right hand you
examine per vaginam and per rectum. Simultaneously you assist
and contribute to the examination by the left hand placed over the
hypogastrium, pressing the parts; the left hand pressing the parts
down upon the right, the right pressing the parts up upon the left.
Making the examination in this way in a woman who is not nervous,
who submits well to the examination, who has not a great quantity
of fat, you can feel the pelvic organs with very great precision. You
can easily, as you will have plenty of opportunities in healthy women
of ascertaining, make the fingers of the two hands meet, for instance,
in front of the uterus; and thus you grasp the bladder between the
two hands. You can place the fingers so as to grasp the uterus and
feel every part of it; and pressing the fingers towards the sacro-iliac
joint on either side, you can, in a great number of cases, seize either
ovary. If you cannot find an ovary, that is not proof that it is
healthy; it is a presumption, however, in favor of its being so, for,
in almost all of the minor diseases of the ovary you can easily seize
it. In a woman with healthy ovaries it is sometimes impossible to
identify these organs in this way. They are then too small, soft, and
mobile. If, however, an ovary is of the size of a walnut, you can
quite easily feel it. This bimanual examination is frequently rendered abortive by the presence of adhesions; and if these are present,
they, as it were, throw a cloud of uncertainty over the examination,
by preventing you from identifying the ovary and feeling it.

I may mention here that at least one eminent author says that the
minor forms of inflammation of the ovary never exist without ulcera-

tion of the cervix ; and that that is a mistake. Were it true, it would
be an extremely valuable indication ; for if there were no ulceration
of the cervix there would be no ovaritis. Such a negative indication
would be of great value, could we rely upon it. This we cannot do.

What do we make out by bimanual examination? We make
out, firstly, tenderness, frequently intense tenderness ; and this may
prevent thorough examination. The woman cannot endure it ; she
shrinks, and will not allow you to proceed further. Be very careful,
therefore, not to elicit more pain than is inevitable, at least until you
are about to finish your examination. You frequently are astonished
to find the presence of this tenderness, for it is often present when the
woman complains of no distinctive ovarian pain. It attracts your
attention in such a case for the first time to the seat of the disease.

The position of the uterus is affected by ovarian swelling. Ovaritis is a disease eminently liable to relapses ; it is liable also to attack
one and the other ovary alternately. In these cases you often observe a
change in the position of the uterus when the ovary becomes enlarged
and swollen. It may be retroverted when the ovary is healthy, and
then be pushed up into a natural situation when the ovary is swollen.
If the ovary becomes increased in size, and, therefore, in weight, you
would naturally expect it should fall down—and so it does ; it becomes what is called prolapsed ; instead of being upon a higher level,
as in health, it lies upon the level of the neck of the womb—sinks
into Douglas's space. This prolapse makes the organ more easily
felt, and in the case of dyspareunia that I read to you, both ovaries
were distinctly felt and both prolapsed, lying down on the level of
the cervix, instead of higher up behind the uterus. This prolapsus
does not invariably occur ; but when it does occur you can easily understand how it should aggravate dyspareunia. A woman with inflamed ovaries in the natural position may not have dyspareunia ;
but a woman with a slight degree of ovaritis, or with only slightly
enlarged ovaries, if they are prolapsed, may have great dyspareunia.
Besides making out the condition of tenderness, you examine the
consistence of the organs. They may be more or less hard or elastic.
Next, you pay attention to the size of the organs. This is a subject
that has been very much disputed. I am satisfied that an inflamed
ovary may increase in size to many times its natural bulk, but not
many times its natural lineal dimensions ; and these two modes of
measuring are often confused. I have seen in a post-mortem a hypertrophied ovary which was as big as a small hen's egg, and

that without anything to be discovered in it except what may be called
areolar hyperplasia, or, in simpler words, increase of fibrous tissue.
In some cases—rare, no doubt—the ovary becomes smaller, and you
have a condition which has been described as that of cirrhosis. In
this cirrhotic disease, the ovary of an otherwise healthy, young, and
vigorous woman has become contracted to a size little larger than that
of a field-bean; and when cut through it is found to consist of
nothing but intensely dense, whitish, fibrous tissue.

The enlargement of an ovary that is the subject of inflammation is
a matter that requires a little more description, because in some cases
that are otherwise naturally classed clinically as cases of ovaritis, you
have merely irritation from the growth in the ovary—not of ovarian
dropsy, but of a thin-walled cyst or cysts, generally filled with a limpid straw-colored fluid, sometimes containing fluid that is grumous,
or at least tinted with blood. There was a case in "Martha" last
Tuesday. The woman came to us in perfect health except with pain
in the region of the left ovary. Her left ovary was about the size of
a small orange, and felt as if it were a tense cyst. I believe, though
I cannot prove it, that it has in it a tense cyst, and it is the tension of
this cyst that is giving her the pain, and nothing that we can do will
do it any good. I anticipate, however, speaking from considerable
experience of such cases, that it will burst and disappear without
giving the woman any more trouble.* If its contents are unfortunately grumous, or mixed with pus, it will probably give the woman
a good deal of trouble—it may be fatal; but I anticipate nothing of
that kind, because she has no symptoms indicating inflammation of
the part. There is very litte tenderness, and no adhesions; I therefore think the pain arises from tenseness of a cyst. This disease is
generally called hydrops folliculi, and the bursting is not rare. Sometimes, instead of one cyst you have several, not a multilocular cystoma, but two or three or more of similar cysts or dropsical follicles.
They are said to be Graafian, but that remains to be proved. It is
certain that some of them are Graafian, because ovules have been found
in them. Several years ago I described this formation and bursting
of ovarian cysts as forming a disease that might be confused with
ovaritis; and since then I have taken great interest in noticing, not
post-mortems—for, as I have told you, the disease we are discussing
is scarcely ever illustrated in post-mortems,—but, what is nearly as

* It did disappear suddenly, and without causing any symptoms whatever.

4

good for the purpose, cases of spaying, an operation which has been
introduced recently into gynæcological practice. If you read recorded
cases of the spaying of women suffering from what was supposed to
be ovaritis, you will find in an astonishing number—certainly in three-
fourths of them—that, on getting hold of the ovary, it was found to
be composed of fragile cysts, whose presence, previous to the operation,
had not been suspected. This complication adds greatly to the difficulty of diagnosis.

The essential part of the diagnosis I have gone over. I shall
now mention a few symptoms which occasionally accompany ovaritis.
These symptoms, no doubt, frequently depend upon corporeal endometritis, which, in a considerable number of cases, accompanies ovaritis.
Corporeal endometritis is inflammation of the mucous membrane of
the cavity of the body of the uterus. This being so, you can easily
understand that in cases of ovaritis you will frequently have either
menorrhagia or prolonged menstruation. Occasionally, however, you
have amenorrhœa; and in such cases there is probably no endometritis.
Women are frequently sterile when suffering from ovaritis; but there
is no invariable connection between the two. Indeed, although it is
probable that there is some connection, it is far from being proved.
I have known typical cases of ovaritis lasting during the childbearing period of life in women persistently fertile.

I must now say a few words on peculiarities of ovaritis; and the
case I am going to read to you will illustrate the tendency of the disease to relapse.

A. O., aged thirty years, married for five years; has had three
children, the last eighteen months ago; no miscarriages; catamenia
natural and regular, the last beginning on December 26th; admitted
January 18th. She complains of pain in the hypogastrium and in
the left iliac region, which has continued since January 1st. On
careful examination the only disease discovered is a swollen, tender,
left ovary. Besides ordinary constitutional treatment, a blister two
inches square was applied above the left inguinal canal, and a course
of half-drachm doses thrice daily of the liquor hydrargyri perchloridi
was instituted. Under this treatment she rapidly improved; but a
week from its commencement—that is, on the 25th—the symptoms
and signs of disease, which were dying away in the left side, began
with some intensity in the right. A similar treatment was used for
this relapse, and with similar good results. This is a remarkably
favorable case. What was the cause of it we do not know, but it was

of very short duration, and there is considerable reason to fear that it will again relapse.

There is a distinction of the disease into follicular inflammation and interstitial or stromatous inflammation, which is of very great importance, but which, for clinical purposes, has as yet no interest. I know no way during life of making any distinction between them. The follicular inflammation is said to be most frequent as a complication of fever or cholera, and to be accompanied by little tendency to the formation of adhesions. The inflammation of the stroma is said to occur chiefly after abortion or lying-in at the full time, and in it you have greater tendency to increase of size and the formation of adhesions.

The presence of adhesions of and around the ovary prevents your being able to diagnose precisely the disease while they persist. It may be exactly diagnosed if you ascertain the condition of the ovary before the adhesions were formed ; or if you ascertain the condition of the ovary after the adhesions disappear; and both of these observations are not rarely made in actual cases. When you have adhesions forming around the ovary you have always a threatening of the formation of perioophoric abscess, which I have told you is not of very rare occurrence. If one ovary is surrounded by adhesions, you frequently have a singularly distinct limitation made out by your examining finger of the condition of the upper part of the cavity of the pelvis as supposed to be divided into four parts. In the case of the inflammation of one ovary, you frequently have one of the posterior quarters of the upper part of the pelvis solidified. You have then a right to suspect that the mass of hardening and adhesions is not the result of metritis, but the result of the inflammation of an organ which is in that part. I shall read to you a case where this was particularly well observed, and where, in accordance with the natural history of the disease, the adhesions spread, so as, in the latter part of the case, to occupy both posterior quarters, that is, the posterior half of the upper part of the pelvic excavation, and, as you might expect, push the uterus somewhat forwards.

S. E., aged thirty-four, married for four years, has had two children and no miscarriages. Last child born about eleven months before admission to " Martha ;" much hæmorrhage at its birth. Catamenia began when she was thirteen years of age : they are always irregular, painful, and profuse. The last period occurred six weeks before admission. About three months ago underwent an operation for fissure

of the anus, which has not relieved her. She now complains of pain
on micturating. Nine days after last confinement she travelled a
hundred miles by train. This brought on pain in the lower part of
the back and in the lower part of the belly, and along the outside of
the right thigh. This pain has been aggravated during the last three
months. Has a little white discharge. There is, on external palpa-
tion, a feeling of fulness over the right side of the brim of the pelvis.
In the same part there is tenderness. Digital vaginal examination
discovers the cervix uteri in a natural situation. The probe discovers a uterus of natural dimensions in its natural situation. The
uterus is fixed, and cannot be discriminated by the finger from a mass
of tender hardness which occupies the right posterior quarter of the
upper part of the pelvic excavation. The tender hardness subsequently increased in bulk, extended across the pelvis to the left side,
and displaced the uterus forwards. She was kept in bed and treated
as I shall presently describe. When, after six weeks, she was dismissed, her symptoms were all much alleviated. The signs of disease
had also disappeared, except an adherent fixed uterus. Here was a
case of subacute ovaritis almost certainly produced by the journey
after the last confinement. The disease had lasted about eleven
months, and during her stay in the hospital it extended from the
right side to the left, but under proper treatment (whether the improvement is to be ascribed to the drug part of the treatment or the
rest and proper hygienic care, I cannot say) the active disease rapidly
disappeared.

Before I conclude, I must say a few words as to the causes and
treatment of this very important disease. Occasionally it is seen as
a consequence of fever, especially typhoid, of cholera, and of rheumatism; and, in close connection with these diseases, it is very frequently a result of the use of alcoholic liquors, even when these are
not taken to excess. At present my impression is that that is the
most frequent cause of the disease; and this view of the causation of
the disease is in the most gratifying manner frequently corroborated,
if not proved, by the cure which follows upon the adoption of tee-
total living. A great mass of cases occurs as a consequence of recent
marriage, suppression of menstruation, abortion, and delivery at
the full time, when there is no evidence of blood-poisoning. In a
certain class of women you have the disease occurring in its most
characteristic form; and it is in young strumpets that the disease is
best studied. There it is a consequence of gonorrhœa. The inflam-

mation extends to the ovaries. It may be chronic for a considerable time, and produce, as its chief annoyance to the patient, slight loss of blood in consequence of the endometritis which, in this case, accompanies it. Then the disease may produce perioophoric adhesions ; and, under proper treatment, you may watch the disappearance of these perioophoric adhesions and the disappearance also of the ovaritis. It is in cases of young strumpets that I have learned most of what I have been describing to you.

Now a few words on the treatment; and I begin by telling you that you will find a great many cases chronic—which is almost a synonym for incurable. I advise you, indeed, in many cases which resist a properly conducted treatment, to give up the attempt at cure. You will only bother your patient, make her a valetudinarian, and do her harm, by further persistence in attempting to cure a disease which proper treatment has failed to remove. The treatment is modified according to the nature of the disease, according to its acuteness. In every case you wish rest ; and, no doubt, the more serious a case is, the more strict should be your injunction as to rest, and in bed. Physiological rest can only be obtained very imperfectly, for the woman must menstruate, and that is an interference with physiological rest. In married women there are other difficulties which do not require to be described. In many cases the use of leeches applied to the neck of the womb, or applied over the inguinal canal, is very valuable. The medicines most relied upon are corrosive sublimate, iodide of potash, and bromide of potash. As leeches are specially useful in the acute cases, so frequent blisters over the inguinal ring or in that region are frequently very valuable in the chronic cases. Lastly, it has of late years frequently been decided to spay women in this disease ; and many cases of the operation are recorded. That operation is still *sub judice*. Most gynæcologists say that it is condemned already, but upon it I reserve my opinion.

V.

PERIMETRITIS AND PARAMETRITIS.

LATELY I told you that procidentia of the womb is purely mechanical, as much so as a dislocation of the shoulder or a hernia. Now, the subject upon which I am to lecture is inflammation in the neighborhood of the womb, and I begin by telling you, what I shall almost immediately afterwards partially contradict, that this disease is purely vital. Although it would be well worth our while, yet it is not at present a proper subject, to enter upon the great question implied in vitalistic doctrine, which has been very extensively discussed in this hospital, a discussion in which very great men have taken part, for instance, Abernethy, Lawrence, and, still more recently, Sir James Paget. I shall only say that I call this inflammatory disease vital, not because I believe in vitalistic doctrines. The whole tendency of my scientific thoughts is against vitalistic doctrines. I believe the time will come when nearly all the diseases of women will be explained by a transcendental physics, including chemistry. But that time is very far distant, and I dismiss this subject, merely remarking that the disease which we are about to discuss is vital in contradistinction to procidentia, which is rudely mechanical.

Time will allow me to dip only superficially into our subject. In medicine and surgery inflammation is the most important of all the morbid processess. So it is in gynæcology. I shall now take two inflammations—one, parametritis; and the other perimetritis. These are two only of many cases of the kind in my wards at the present time. Indeed, inflammations of the genital organs are the stock-in-trade of the gynæcologist. There is no doubt they are by far the most prevalent and the most important uterine diseases. In the womb, as in all other organs, the great causes of inflammation are injury and cold. It is difficult to decide in most cases which is the more potent; frequently they are combined, and you can easily understand the combination and the commonness of the disease when I tell you that these inflammations so frequently follow menstruation, miscarriage, and delivery at full term. You can readily imagine the tremendous influence of exposure to cold after the injuries and bruises implied in the last of these processes.

The next point upon which I must say a few words, in order to lead the way to the cases before us, is the nomenclature. This, if I had time to give it fully, would be to a great extent the history of our knowledge of the diseases. When we speak of inflammation localized in individual organs, we speak of inflammation of the womb, or metritis; inflammation of the tubes, or salpingitis (a disease almost unknown—our knowledge of which is chiefly derived from the dissecting-room); and lastly, inflammation of the ovaries, or ovaritis. The subjects of my lecture are perimetritis and parametritis, and you must not suppose that they are separate from the three diseases I have just mentioned. They are merely so on account of our too frequent ignorance of their origin. They are not separated theoretically, but for the purposes of clinical teaching. Suppose you were to put a tangle tent into the uterus of a woman without any precaution, leaving it there for days; in all probability it would give rise to perimetritis or parametritis. There would be pain and tenderness, etc., and you diagnose one of these diseases. It is not, however, an inflammation around the womb merely, but also inflammation of the womb itself, although the only tangible evidence you can get is of peri- or parametritis, not of metritis proper. This nomenclature, therefore, is chiefly nosological or practical, and not pathological or scientific. Perimetritis is very frequently spoken of as pelvic peritonitis. A very common term for parametritis is pelvic cellulitis. I think it an objectionable name, although it is very much used. I have no time for stating my reasons for this objection.

The next point, before coming to the cases themselves, is to beg you to dismiss from your minds two errors in regard to these diseases; and, if they have not entered your minds already, to keep them away. The first of these is the notion that these three diseases—metritis, salpingitis, and ovaritis—are rare, especially in unmarried women, or women apart from the accidents of pregnancy. This was, however, a very prevalent idea, and a very erroneous one. The causes of error are very easily found. First, is the neglect of the proper method of study of these diseases. One of the most respected teachers in London long ago described a disease which was for a long time known, or supposed to be known—namely, irritable uterus. This is embalmed in the minds of all the old doctors now living. We might as well talk of an irritable nose, or an irritable tongue. I never saw an irritable uterus of the kind referred to. I mean no disrespect to the great Dr. Gooch: his day is past, as mine will be ere long. About thirty years ago young ladies frequently had spinal irritation. What that was I do not know. This expression lived in my day; it is now as much dead as irritable uterus.

Great improvements now arise from *touching* everything and *looking* at everything. By these means we are enabled to recognize metritis and ovaritis as far from rare apart from pregnancy. But we recognize in unmarried women, and in women apart from the accidents of pregnancy, metritis or ovaritis *proper*, comparatively seldom.

We find the evidence of these diseases having originally existed in perimetritis and parametritis. This is what you must understand. In our much-used textbooks, at one part we find metritis and ovaritis, and quite in another part parametritis or pelvic cellulitis, as if it were a totally distinct disease, which it is not. It would be the same error to describe a bubo as an abscess, or suppuration of the cellular tissue, and nothing more; whereas it is originally an inflamed gland. Suppose we pass a catheter into the urethra of a man suffering from a stricture; this may give rise to a perineal abscess. Inflammation of the strictured portion spreads to the surrounding parts. So it is with inflammation of the uterus. The effusion and suppuration take place in the neighborhood. The whole heart may be inflamed; but the outer or inner membrane, the pericardium or endocardium, shows the inflammation the most.

Again, up till lately we almost never heard of perimetritis or pelvic peritonitis, only of parametritis, or pelvic cellulitis, or of pelvic abscess. All these inflammations or abscesses were supposed to be in the cellular tissue. Perimetritis was almost known.

This great improvement—the discovery of the frequency of perimetritis—we owe to Bernutz, a Parisian physician still living. This increase of knowledge and of our beneficent powers was not only made good for perimetritis, but also for a closely related disease, hæmatocele. In all my early life I never heard of such a disease; and when I heard of it, it was only as an effusion of blood into the cellular tissue, like a great black eye or thrombus. Bernutz showed that the great majority of *large* and *grave* hæmatoceles and pelvic abscesses are not in the cellular tissue, but in the peritoneum. Bear this in mind. I do not say the great majority of hæmatoceles and pelvic abscesses, though I have been represented as having said so; but the great majority of *grave* and *large* hæmatoceles.

The last point I shall mention in this connection is the commencement of our knowledge of induration around the womb. The man who, long before Bernutz, made this discovery was Doherty, who afterwards became Professor of Midwifery in Galway, now dead. This was the foundation of future progress. Doherty knew nothing of the distinction between perimetritis and parametritis. He merely recognized pelvic indurations. "Hard as a board," were the words he used, and they are still employed. He knew that these were inflammations, and not necessarily abscesses. This is a point of great importance in the pathology of this part of the body.

Doherty began a series of investigations which have ended in this, that there may be two kinds or degrees of cellulitis or parametritis. The first is sometimes called phlegmon, to distinguish it from suppuration, or abscess. The term inflammatory induration is generally applied to the former. Suppose you have a little boil on the hip, it will be surrounded by an extensive inflammatory induration, perhaps as big as a saucer. This is the same kind of change as takes place around the womb, from inflammation which begins in its structure. If an intra-uterine pessary be inserted, without proper care being taken, and should the patient be seen a week afterwards, probably, instead of finding everything soft and movable around the cervix, a tender hardness may be found around the womb, to use Doherty's words, "as hard as a board;"—that may be parametritis. That this may also be perimetritis was the great discovery of Bernutz.

The lumps produced by perimetric adhesions were generally mistaken till his time. I remember a case diagnosed as a fibrous tumor of the uterus, a rounded hard mass, as big as a child's head, above the brim of the pelvis, very slightly tender, fixing the uterus. The young lady died, and at the post-mortem examination it was found that there was no fibrous tumor at all. It was adhesive perimetritis —a packet of coherent intestines, which formed a hard mass, and had led to the deception of eminent and experienced gynæcologists.

When perimetritis occurs, generally the ovaries and intestines and broad ligaments and parietal pelvic peritoneum become glued together, forming a hard tumor. All this perimetric swelling may ere long be dissipated like snow off the streets, just as often happens with parametric phlegmon.

We now come to two cases, and with these I must be brief. The first is a very interesting one. I will go over the most important points in it.

A. M., confined naturally ; seven days afterwards was unable to pass water, and had a shivering fit. This was the commencement of the disease, and it occurred five months before her admission to the ward. It was a case of parametritis. Observe what a chronic disease this may be, and you will see it is liable to relapses. On the tenth day the patient became better, and left her bed, but she never got rid of the pain in the left iliac region, which came with the shivering. In a fortnight's time she was worse than ever ; then again she got a little better, but subsequently her symptoms became more severe, and, when she came to the ward, what did we find ? The uterus was displaced a little to the right side,—you will remember the pain was in the left,— and from the neck of the womb to the wall of the pelvis on the left side the roof of the vagina was " hard as a board," the uterus fixed. Why did we call this parametritis ? Chiefly because we felt no mass, simply a hard, tender surface. If it had been perimetritis we should have felt a mass by bimanual examination, a packet of intestines and tube, etc., matted together. What we did find was such hardness as occurs around a boil or inflamed gland. Again, had it been perimetritis there would have almost certainly been tenderness on pressure in the left groin ; and, lastly, if it had been perimetritis there would not have been a flat surface extending to the bone, but a somewhat shaped tumor in the roof of the vagina. One evidence of the correctness of the diagnosis is that if the woman be examined to-day, it will be found that the induration has almost entirely disappeared, and that the cervix is close to the pelvis, as if the uterus was adherent to the left sacro-iliac synchondrosis. By-and-by, we expect, it will get again mobile.

Now, a few words about a still more important case,—a woman who for a long time swam for her life, having had an attack of py-aemia in the course of her recovery from a perimetric abscess. This was of the most frequent kind, sometimes called retro-uterine, because it has the same relation to the uterus as a hæmatocele, such as has been described as retro-uterine hæmatocele by Nélaton (and it is often difficult to diagnose the one from the other).

In the case of which we are speaking there was an enormous mass which pushed down into Douglas's space, driving the uterus against the symphysis pubis and extending upwards as far as the umbilicus, displacing the bowels, so that it felt like a gravid uterus at the sixth month. The patient was a charwoman. She became suddenly ill, and had to leave her work. The same night she felt intense pain in the left iliac region. This was seven weeks before she came into the hospital. There was no rigor. A fortnight after the attack a lump was felt in the abdomen, which gradually increased till it reached the enormous size described. The diagnosis was arrived at from its size, its position displacing the uterus and bowels, and its great tenderness. The abscess burst into the bladder, and torrents of pus passed through it; pyæmia then occurred. She became dangerously ill, but is now nearly well. The lump has entirely disappeared.

What was the treatment of both cases?—Antiphlogistic! The same, without entering on minutiæ, as for inflammation of any other region. The most important element, however, in the treatment was lying in bed, a remedy of more value in such cases than all other medicines or drugs put together. This was all that was prescribed in the first case. In the second, surgical means would have been adopted had not the abscess burst into the bladder. The treatment, therefore, was lying in bed. The proper use of the knife is extremely restricted, so much so that many of the greatest gynæcologists say that these abscesses never should be opened. This, however, I believe to be a mistake. Had the disease lasted longer in the second case without discharge of pus, I should certainly have proceeded to evacuate it by an incision per vaginam.

VI.

ON KINDS OF PERIMETRITIS.

THE diseases to be now considered are various forms of internal inflammation. That pathological condition is the cause of the most frequent, and therefore—and for other reasons—the most important diseases of women. Inflammations not of distinct organs, as of the uterus or of the ovaries, are divided into two sets,—perimetric and parametric. To-day we consider inflammations in the former category. When I say "inflammations not of special organs," I do not wish you to understand that inflammation of special organs, as ovaritis or metritis, has nothing to do with the diseases under consideration; for, although we speak, for example, of metritis and of perimetritis separately, yet the metritis, or it may be, ovaritis, is frequently the cause of the perimetritis, and error as to the frequency of inflammation of special female organs arises from neglecting this circumstance.

There are three kinds of perimetritis,—adhesive, serous, purulent. Of these three the purulent is the most important, including, as it does, a large number of what are called pelvic abscesses; but of the purulent form I have no time to say anything at present. There may be another kind of perimetritis characterized by dryness and slight roughness of the peritoneum; but, from the deep position of Douglas's space in the body, this form of inflammation has, so far as I know, never been recognized in this situation. It is common in cases of uterine fibroid and in cases of ovarian dropsy, and is easily made out by friction being both felt and heard. It often lasts for a long time, and does by no means always end in producing adhesions.

Adhesive perimetritis is almost certainly second in point of frequency among the diseases of women, the first position being held by uterine cervical catarrh. In post-mortem examinations of women no pathological condition is more frequently discovered than adhesions between the internal genital organs and neighboring parts, especially about the ovary. The disease is generally characterized by very little, or by complete absence of, pain; it is generally, not always, narrowly limited in extent; and generally, so far as the life of the patient is concerned, it is of little importance. Persistent adhesions are sometimes the cause of aching or more decided pain, as John Hunter knew; but then persistence is a somewhat rare occurrence. Adhesions in this situation are gradually worn away and removed just as they are when in the pleura or in the pericardium; and about the womb the same filmy or stringy adhesions, sometimes of great length, are not rarely observed, just as they are observed in post-mortems in the pleural cavity or in the pericardial. At length all these adhesions disappear. The upward and downward movements of the viscera rub them all away. They persist longest and may be never removed about the ovary, and when persistent there they connect the ovary with the neighboring tube and broad ligament; and you can easily understand why they are not removed here, for the mechanism of their removal is absent, there being produced by inspiration and expiration no movement of the ovary upon the parts to which it remains persistently adherent. It is only in rare cases that uterine adhesions remain, either getting so organized as to resist the mechanism of their removal, or being so perfectly fixed as never to be subjected to it. An instructive illustration of this disease in its earliest stage was lately seen in the post-mortem theatre. The case was one of malignant disease of the cavity of the uterus, and solution of perchloride of iron had been injected into it. The woman died within two days afterwards. She had no complaint of pain in the region of the womb, but she had perimetritis—the characteristic perimetritis confined to the peritoneum of Douglas's space. The report of the autopsy says: "The posterior surface of the uterus and Douglas's space were injected and covered with a thin layer of recent lymph." Before I leave this kind of perimetritis I must remark upon its recent and insufficient recognition, and this arises chiefly from the circumstance that the professional mind connects with peritonitis high fever and intense pain. An adhesive perimetritis frequently occurs without any symptoms at all—indeed, generally. It produces fixation of organs and the feeling of solid lumps in the pelvis; and these feelings have hitherto proved a fertile source of error.

Before I leave this subject I must say a few words regarding a case which has just been dismissed from "Martha." In this case the adhesive perimetritis was produced by the use of pessaries—a not unfrequent cause of the disease. The case illustrates well the physical conditions and the rapidity with which evidence of the disease may disappear. At the end of the case you will observe the contrast with the description of the beginning, there being only left a little tautness on one side, no doubt the result of adhesion still persistent in that situation. You will remark also how, as improvement went on, a lobulated feeling supplanted the previous more uniform surface, the disappearance of the diffused inflammatory swelling allowing the forms of the organs to be felt.

Mrs. C., aged twenty-four, married for three years, had a child a year before admission, and a miscarriage seven months thereafter. Menstruation is copious, painful, and lasts a week. Micturition has been painful for the last fortnight. Feels and looks ill. Temperature 99.6°. Complains of pain and tenderness in the lowest part of the abdomen, chiefly on the left side. This she has had more or less for three years, and she has worn a large number of instruments for "bent womb." Latterly she has been so much worse as to seek refuge in the hospital. Per hypogastrium tender hardness can be felt occupying the brim of the pelvis. The cervix uteri is in its natural situation, and all around it, especially posteriorly, is dense, uniform, tender hardness extending to near the bony brim, and producing fixation of all the organs and parts in this situation. The uterus is natural in position, and there is no special tenderness of its interior. She was ordered to have moderate diet, to be confined to bed, to have hot cataplasms to the hypogastrium, and to have the bowels regulated. Under this treatment she gradually improved, losing all pain and tenderness. After nine days' confinement in bed, examination discovered great diminution of the tenderness of the hard and fixed parts; there was a lobulated feeling in the hardness, the organs being felt through it; and the uterus was nearer to the sacrum. After sixteen days' confinement in bed, the hardness was no longer to be felt per hypogastrium. Per vaginam the uterus could be made out, somewhat anteverted; its cervix adjoining the upper part of the sacrum. It is easily displaceable, and no adhesion can be made out except on the right side, where there is tautness produced by moving the uterus from it. She now feels and looks well.

I now come to consider Serous Perimetritis. This form of the disease is very uncommon, at least in a well-marked form. In an imperfect clinical form it is frequently observed in the post-mortems of cases where there has been peritonitis, there being found, under such circumstances, adhesions at one part and little collections of serum at another. Such collections may be numerous, and some of them may be of pus, not of serum. The serous perimetritis that I am now speaking of resembles purulent perimetritis or pelvic abscess in every respect except the slighter degree of the symptoms of suppuration. Its special character is only, so far as I know, recognizable by evacuating the cavity. This is done according to the same surgical plans as in cases of purulent perimetritis or pelvic abscess; and when it is done, the peculiar nature of the disease is discovered, serum being withdrawn instead of pus. Analogous collections of serum are more frequently observed in the pleura than in the hypogastrium; in the latter situation they are undoubtedly rare. In the first case which I shall read to you, the disease was characteristically retro-uterine, the tumor filling the pelvis, and bulging downwards in a manner often compared to a stage of progress of the fœtal head, filling the pelvis. This retro-uterine position is common in hæmatoceles. In a well-marked case you have to press the finger between the swelling and the symphysis pubis in order to reach the cervix uteri, which is displaced upwards. So it was in the case that I shall read. This forcing of the finger you have also to do in some uterine and ovarian tumors, and in retroversion of the gravid uterus; and it is well to keep these facts in mind. The case I am about to read was supposed to be an abscess, and it was resolved to open it behind the cervix uteri. The finger was applied to feel if any artery was pulsating at the point of selection in the mesial line. This point is selected in order to insure entering the expanded Douglas's space. Here, in a healthy woman, the peritoneum comes in apposition with the roof of the vagina; but I would not warrant that any of you would puncture Douglas's space with any certainty in a healthy woman, for the extent of vaginal roof covered by peritoneum is then very small; but when Douglas's space is expanded by being filled with serum, pus, or blood, which are detained there by adhesions formed above, inclosing the serum, and pushing it downwards into the pelvis to a greater or less degree, then there is no difficulty. If you are sure of the nature of your case, and of the need for interference, you plunge your knife into the point of selection, and the fluid flows. Now for the first case.

Mrs. B., married four years, had one child four months ago; no miscarriages. About a month after her confinement she began to have pain in the hypogastrium, and shortly afterwards pain and difficulty in micturition, and constipation. On April 4th she was admitted to "Martha." Her bowels then had not been opened for a week. They acted freely after a dose of castor oil. Temperature in morning 100.8°, in evening 101.6°. The lower part of the abdomen is occupied by a hardness, somewhat tender, nearly of the shape and size of a four-months' gravid uterus, but not of the same feeling. There is comparative dulness on percussion over it. The pelvis is occupied by a globular elastic mass. The cervix uteri is with difficulty reached by pressing the finger between this mass and the pubes. The uterus is above the pubes. The probe passed into it enters two inches and a half, and it is deflected to the right side, the fundus being situated about four inches above the right Poupart's ligament. The tumor in the pelvis makes the perinæum bulge, and the rectum at the anus is partially everted. The tumor was opened per vaginam by incision in the mesial line, when about a pint of slightly turbid serous fluid escaped, and the hypogastric swelling disappeared. Next day the temperature was 99.8°, and afterwards soon normal. Slight blood-stained discharge went on for a few days. Ten days after the opening of the collection there was to be found only induration behind the uterus, which was fixed. She left the hospital feeling quite well.

Local peritonitis of any kind, or perimetritis, is not a rare complication of a uterine fibroid or of a small ovarian tumor; and the next case which I shall read to you is an illustration of this disease. The case, I admit, is not so clear as the one just related to you, about which there could be no doubt; but I have scarcely any hesitation as to the cause of the conditions in the large serous collection, whose history I have now to read. There was never any evidence of the existence of a sac inclosing the serum; indeed, the whole clinical evidence tallies only with the presence of a serous perimetritis of an extensive kind. The basis of the disease was, no doubt, a tumor of the pelvic organs, which was not discovered until late in the history of the case. You will observe that during its progress the urine suddenly became albuminous, and bloody stools were passed. These occurrences were simultaneous with diminution of the serous accumulation, and were, at the time, supposed to arise from discharge of the serum through the bladder and through the rectum. This is, however, only a supposition. I am indebted to Dr. Church for this case, having been consulted by him repeatedly in regard to it.*

L. M., a virgin, aged eighteen, healthy-looking and well nourished,

---

* The subsequent history of the case confirmed the opinion in the text.

was under treatment three months before admission for pain in her right side. Nine days before admission it returned, and has been getting worse since, and her attention has been drawn to a swelling in the lower belly. Pulse 120; temperature 101°. The lower half of the abdomen is excessively tender, especially on the right side, and is occupied by a large tumor, which protrudes in the middle. The tumor is dull on percussion, with some resonance in the flanks, especially in the left. It rises to an inch above the umbilicus. The uterus is found to be low down, its cervix pointing forwards. The roof of the vagina is not hardened, and the tender swelling above can just be felt through it. The catamenia are regular, profuse, not painful. Micturition very difficult. The day after admission she was tapped by Mr. T. Smith, and a pint of clear brownish serum drawn off; it became nearly solid on heating. This operation relieved her pain. The day after, pulse 100; temperature 99°. Eight days after the operation, pain and fever returned, but soon subsided. On the ninth day after the operation, the urine contained pus and albumen (half). In a week afterwards the urine was again normal. Two days after, pus and albumen were again found in the urine. She had bloody stools, with diarrhœa for a day. All this time the tumor was gradually diminishing. Seventeen days after the tapping it was found that the abdominal tumor was gone. Dulness in the hypogastrium extended two inches upwards from right Poupart's ligament. The uterus is high up, and far back in the pelvis. Above it, in front of it, on its right side, and just accessible by the tip of the examining finger, is a rounded tumor of the size of an orange, displaceable, but not freely mobile nor presenting distinct indications of connection with the uterus. She was discharged in good health.

I now come to the last subject of this lecture,—Remote Perimetritis. This kind is not in the same category with adhesive, serous, and purulent perimetritis. Each of these kinds may be remote. There are remote inflammations of serous membrane, well known in female pathology, which are produced through or in connection with a constitutional affection, as septicæmia. The remote perimetritis we are now considering has no such origin or history. It is analogous to the remote parametritis, upon which I hope to lecture soon, and whose history is better known than that of remote perimetritis. In remote perimetritis the inflammation was at one time continuous with the pelvic peritoneal inflammation. In remote perimetritis the pelvic peritoneal inflammation may have disappeared, while the inflammation

persists in a remote region; or the remote inflammation may coexist
with the persistent pelvic peritonitis.   The best example I have ever
seen of the remote perimetritis, where all inflammation of the uterus
and its neighborhood had disappeared, occurred in one of the clinical
wards of the Royal Infirmary of Edinburgh, and I was called to it
by the late Professor Laycock.   The history of that case showed distinctly that the peritoneal inflammation which persisted had originally
been part of an extension of a perimetritis after delivery.   When I
saw it the uterus was mobile, and the roof of the pelvis was soft and
not tender.   The only disease was a rounded swelling containing
fluid, situated below the navel, and to the left side, and which produced the constitutional symptoms of suppuration.   The prognosis of
the case was favorable, and it ended as had been predicted.   The abscess burst into the intestinal canal and suddenly disappeared, leaving
behind it only local hardness.

I shall conclude my lecture by giving you an example of remote
perimetritis, coming on many weeks after an unfortunate confinement
and imperfect recovery, where the inflammation extended to the remoteness of the umbilicus, but maintained its continuity with similar
inflammation in the roof of the pelvis.   This case is remarkable for
many reasons.   It is rare; it occurred long after the liability to perimetric attacks following delivery had ceased to be dreaded.   It caused
severe constitutional symptoms, but the local symptoms were of the
slightest kind; yet the physical examination discovered abundant
evidence of the local disease.   The patient was the wife of a physician, and was very carefully watched, and the history is very instructive.   The evidence of a large tumor excited great alarm, from the
difficulty of being satisfied as to its exact nature.   I never doubted
as to what it was; and the history of its origin and of its complete
disappearance leaves no room for entertaining any view of it other
than that it was a case of remote adhesive perimetritis.   The tumor
could be handled freely, and in every part of it there was resonance
on percussion.   I here give you only a sketch of the case.

Mrs. M., aged thirty-six, has had, since her marriage at twenty
years of age, nine children and two miscarriages, and was confined
on May 25th of her tenth child.   Had hæmorrhage an hour and a
half after the birth of her second child; labor natural.   Tenth child
weighed ten pounds and a half.   After the birth of the child a
drachm of ergot was administered.   Fever supervened on the second
day, and was subdued by the liberal use of salicylic acid, which at

the same time produced very painful symptoms. On the tenth day after delivery I was called, and found her feverish and suffering from occasional nausea. The uterus was large, rising nearly to the umbilicus, but not tender; the lochia not offensive. Ergot was administered, and the following day a rounded decomposing clot of blood as big as an orange was expelled. The uterus gradually assumed its proper dimensions. She suckled her child till near the end of the second month. About this time she went out driving more than once, but her temperature never descended to its normal condition. Then she was suddenly seized with the perimetritic disease. It was announced by giddiness, and severe and persistent bilious vomiting, slight acceleration of pulse; no further rise of temperature, no abdominal pain or tenderness. Some diarrhœa, weakness, prostration, and emaciation came on and gradually increased; and Professor Gairdner, who saw her, considered her disease to be an indistinctly defined swelling in the hypogastric region, the result of an inflammation having its origin in the pelvis. The swelling was scarcely tender. I again visited her on the sixty-seventh day after her confinement. The whole lower half of the abdomen was occupied by an ill-defined, moderately hard, slightly tender swelling, not dull on percussion, evidently formed by the matting together of the pelvic viscera and superjacent intestines as far as the level of the umbilicus. The uterus was fixed, and slightly tender hardness surrounded it. She was kept in bed, and the abdomen was anointed with iodized oil. She slowly recovered. Before six weeks had passed from my visit, the temperature had resumed and maintained a normal condition, and she began to put on flesh. After some more weeks no trace of the abdominal tumor could be discovered on the most careful examination. Since then she had another child without any trouble or alarm.

VII.

FORMS OF PARAMETRITIS.

PARAMETRITIS, or inflammation of the cellular tissue around or in connection with the womb, is one of the most important subjects in gynæcology. I can only, in accordance with the cases that I have to consider, go over a very small part of this great subject. Parametritis may begin and end during pregnancy, and give rise to great difficulties in diagnosis when it is pelvic. Of this I have seen an example. But in pregnancy it is very rare. It is characteristically a disease of the puerperal and of the unimpregnated states. There is a kind of parametritis which I do not consider at all, and which is observed in cases of septicæmia or pyæmia, or what are ordinarily called cases of puerperal fever. This parametritis is erysipelatous in its nature ; it is diffuse and it is not in its general characters like ordinary inflammation. There are pathologists of eminence who regard all kinds of parametritis as essentially the same, differing only in degree. In the meantime, at least, I do not hold that view.

The kinds of parametritis are phlegmon, abscess, gangrene ; and these again may occur in different forms. You may have a chronic parametritis, a chronic phlegmon, ending in the production of indurations, which, when cut into, present a hard, dense, fibrous structure, the interstices of the fibrous tissue being filled up with fat. These chronic hard masses are most frequently observed at one or both sides of the uterus. They sometimes atrophy, like fibrous tissue, the product of inflammation in other situations, and may produce hydronephrosis by compression of the one or other ureter ; as well as fixation of the uterus in abnormal positions.

Parametric phlegmon is a common disease. Like all forms of parametritis, it is most frequent in the close neighborhood of the uterus, and especially on either side of the cervix, where there is plenty of cellular tissue to be the subject of the disease. But a parametric phlegmon, as I shall presently explain to you, may be remote ; any form of parametritis may be remote ; and my lecture to-day is chiefly devoted to remote parametric abscess. Before I leave the subject of parametric phlegmon I shall still further explain to you what it is. It is that kind of tender swelling and hardening of cellular tissue around the womb, or in connection somehow or other with the womb, similar to what you see elsewhere around an inflamed gland or around a carbuncle; and which, when the inflammation ceases, or when the carbuncle disappears, melts away, without suppurating. That is called, in the case of the uterus, a parametric phlegmon; it is common. An ovariotomist occasionally, but very rarely, sees this. In the post-mortem theatre it is extremely rare distinctly to see it, for it is not a fatal disease. An eminent ovariotomist, describing parametritis, not in the pelvis, not forming a pelvic tumor to be easily reached by the finger passed into the vagina, but describing a parametric phlegmon around the brim of the pelvis, uses these words: " Looked at from above, the swelling of the pelvic soft parts was such that three fingers only could be pushed into the cavity, instead of the whole hand with a big sponge in it as is usual." Here was a parametric inflammation in a woman who was subjected to ovariotomy, and who never had an abscess; but all this great parametric phlegmonous swelling disappeared, just as the great hard swelling around a carbuncle disappears, without any suppuration.

Parametric abscess, which is the chief subject of my lecture, may be a degenerated or an advanced phlegmon, or the case may be one which has run on to suppuration at once; and I shall give you examples of these courses in different histories. But I have said that it is chiefly remote parametric abscess that I am to lecture on, and this leads me to consider the subject of the extension of parametric inflammation—a most important subject in practice. After delivery, for example, a woman's uterus gets inflamed, or after a surgical operation upon the uterus, the organ gets inflamed; all the cellular tissue around it, especially on each side of it, may become swollen, hard, and tender. The inflammation extends, and the directions of its extension are very important for you to know. The most frequent is along the pelvic brim towards the psoas muscle, and further on towards the kidney or even around the kidney. It does not extend downwards towards the vulva or ano-perineal region.

Pelvic abscess, as distinguished from pelvic phlegmon, extends frequently downwards into the thigh, and the abscess finds its way among the great muscles of the limb, sometimes even far down. An abscess may spread into the iliac fossa. There is no direction in which it may not spread; but I believe that the phlegmon, as distinguished from abscess, rarely if ever spreads, except, as I told you, upwards along the psoas to the kidney, or forwards, which I did not tell you, along the round ligament to the inguinal canal. Abscess does not extend downwards along the vagina towards the vulva and ano-perineal region, its progress in this direction being probably prevented by the pelvic fascia. Now, these spreadings may be either mechanical, or what we may call, meantime, vital. The mechanical form of spreading some ingenious pathologists have very wisely tried to elucidate by experiment, injecting fluids into the parametric cellular tissue, and observing how the fluid goes, to find out if that accounts for the spreading of abscess or of phlegmon. These experiments have not come to much, and I shall satisfy myself to-day by giving you examples of the two kinds of spreading—examples, which, I think, prove that there are two kinds. The spreading of inflammation so as to affect the inguinal canal, and the disappearance of inflammation everywhere else except in the inguinal canal, is a spreading of inflammation which I in the meantime call vital, not mechanical. No experiment has ever illustrated the running of parametric fluids along the course of the round ligament. Besides, phlegmon, without abscess, may advance into the cellular tissue in the groin, and that while elsewhere, except in the inguinal canal, the inflammation is resolved. I take, as an illustration of the mechanical form of spreading, the advance of matter down into the thigh. That appears to me to be purely mechanical; and one reason for thinking so is that I have never observed the spreading of the phlegmon or mere inflammation in this direction. I have only observed the mechanical forcing of an abscess down into the thigh. That is purely mechanical. The matter not finding vent otherwise, in some cases is urged down into the thigh, and sometimes passes round the head of the femur and gets into the hip-joint, and so produces very dangerous conditions. Sometimes it passes through the great sciatic notch into the hip. The freer or easier progress of pus in a newly found route sometimes explains the diminution of an abscess while yet not opened, either spontaneously or artificially. Of this, an illustration occurred lately in "Martha." A psoas abscess following delivery was expected, after much delay, to open spontaneously in the groin. Consultation was held with Mr. Smith to consider the desirability of immediate evacuation, the constitutional symptoms of suppuration being urgent; but delay was decided upon, in consequence of the depth and difficulty of reaching the abscess. Presently the prominent part of the abscess grew smaller, and this change almost shook our belief in the diagnosis; but the change was soon explained by the appearance of signs of the extension of the abscess through the obturator foramen to the upper and inner part of the thigh.

I must pass now from the consideration of these two forms of spreading—of inflammation and of abscess—the mechanical form and the vital form; and I will give you an illustration, that occurred in "Martha" not long ago, of inguinal parametritis.

S. S., aged twenty-five, has had three children, the last born four weeks and four days before her admission into the hospital. She got on well till a week after her confinement, when she had a prolonged rigor. Shortly after this a lump began to appear in her right groin. At first it gave her little trouble, and a fortnight after her confinement she left the hospital. Two or three days after this exposure the lump in the groin began to increase and be very painful. On admission there is found, extending from near the right anterior superior spine to the body of the left pubic bone, a prominent pear-shaped mass, the smaller end being near the iliac spine. It feels as if filled with fluid. At its broadest it is two inches and a half, measuring upwards from Poupart's ligament. The uterus, and all that can be felt per vaginam, is soft, mobile, and healthy. Only, in the right anterior quarter of the pelvic brim, pressing high up, the finger meets fulness, produced by the above-described swelling as it overhangs the horizontal ramus of the right pubic bone. The abscess was opened with antiseptic precautions and dressed. The finger, introduced into the cavity of the abscess, found it to be limited to the region of the external swelling. The discharge soon dried up, and she rapidly recovered.

This case of inguinal parametritis is an example of spreading to the inguinal canal, indisputably vital. When she was examined, on admission to the hospital, there was no trace of any disease about her womb. The disease began, as you observe, a week after delivery, and remained for a considerable time evidently in the state of a phlegmon; and then she went out. The effect of the exposure was to increase the inflammation and produce suppuration. It was an inguinal parametric abscess, dependent originally upon uterine inflammation, which, when we saw her, had entirely disappeared—a remote form of parametritis.

I have told you that remote parametritis may affect the region of the psoas muscle, or may affect the uterus. The case which I am to read to you is an illustration of remote parametric abscess which maintained to the last some connection with the uterus. It is an example of a kind of disease that is sometimes very puzzling, as I shall presently explain to you. The abscess did not mature and burst till about seven months after the woman's confinement; so chronic was it.

Mrs. R. S., aged twenty-seven, has been married for seven years, and had four children; no miscarriages. Her last confinement was six months before admission to "Martha." It went off easily; but three days afterwards she had shiverings, and has not been well ever since. Now she is exhausted and emaciated, has a quick pulse and a high temperature, both of which rise in the evening and are accompanied by ordinary hectic symptoms. Catamenia have not returned since confinement. She lies on her right side to save pain. The right thigh retracted fully forty-five degrees. Micturition frequent and painful; urine acid, and contains a very little albumen; complains of great pain and tenderness in the right inguinal and hypogastric regions, and shooting down the right leg. These pains have been getting worse ever since her confinement. The lower part of the right side of the abdomen is occupied by a rounded tender hardness, dull on percussion, extending from the right side of the right pubic bone in the direction of Poupart's ligament, and upwards into the right flank, where its distinctness is lost. It enlarges as it approaches the crest of the ilium. The uterus is in its natural situation, but fixed. The right side of the pelvis, but only as high as the finger can reach, is occupied by a tender hardness. She was kept in bed, well fed and cared for, and constantly poulticed. Three weeks after admission, and about seven months after her confinement, relief came by discharge of pus through the already irritated bladder. The tender swelling immediately diminished. The pus flowed freely. The retraction of leg gradually yielded. A fortnight after the discharge of pus began she could keep her leg extended. In a few days more the lump in the right side could not be discovered. The discharge of pus ceased. Two months after admission, and between five and six weeks from the commencement of the evacuation of the abscess, she was discharged quite well.

Here is a case of psoas and iliac abscess, the inflammation beginning three days after delivery, and relief not coming till about seven months afterwards. In this case the abscess was not in the pelvis, the most common seat of the abscess; there was no intra-pelvic dis-

case. The abscess was remote, but maintained its connection with
the uterus—came down as far as the brim of the pelvis and to the
uterus, so as to hold it fixed.

Some of the most important symptoms I shall now describe to you.
She lay upon the affected side, and the case I am to read next to you
lay in the same way upon the affected side, or inclined to the affected
side. The decubitus is sometimes on the healthy side. Her thigh
was retracted, and in this woman it was retracted to about half a
right angle, measuring from the position in standing or extension.
The amount of retraction varies. Extension of the retracted thigh
is almost impossible; it can be done, but it causes so much pain that,
unless under the influence of chloroform, it would be cruelty to try
to do it. It disappears soon after the abscess is discharged, and
before it is completely healed. In my opinion, it is not the result of
neuritis, because there is no special pain along the course of the
nerves, because there is no pain when the leg is at rest, and because
it comes and goes with the disease; whereas neuritis might, as in a
case we have had in "Martha" lately, continue long after the origi-
nal disease had gone. There are cases where you have evidence of
neuritis, but in the meantime it is my opinion that most cases depend
upon inflammation or destruction of the muscle—the psoas and the
iliacus. Another great symptom is the emaciation and haggard
appearance, which in some of these cases cannot be exaggerated, leading bystanders to form an unfavorable prognosis, in which you would
also join if you did not know what this case, and others that you
have seen, illustrates, that this haggard, emaciated deathlike condition disappears with extraordinary rapidity when the abscess bursts
and when the case begins to improve. Now, in many cases of this
kind, mistakes are made, and the mistakes arise from neglecting the
remoteness of the inflammation, the inflammation sometimes being
confined to the region of the psoas or the region of the kidney, while
the region of the uterus, where the inflammation began, has become
perfectly healthy—that is the remoteness of the disease. Mistake
also arises from the difficulty of understanding that the disease can
be so chronic as it is in some cases, and as it was in the woman whose
history I have just read, in which the abscess never pointed, and
burst through the bladder seven months after confinement.

Another source of error is, that in some cases there is no great pain
or tenderness in the region affected—the region of the psoas muscle;
and frequently, as in the case I am presently to read to you, nothing

to be felt. The inflammation produces a flat swelling, which is covered by the bowels, and you do not feel anything through the bowels. The physician cannot find anything like a distinct abscess, even if he happens to expect it. I remember well being called by two excellent physicians to see a case of this kind. It was many months after delivery when I saw the woman for the first time. The physicians had given up the idea that there was a psoas abscess or any abscess following delivery, for there was nothing to be felt. The woman was lying in bed, without much suffering, but quite helpless from the retraction of her thigh, and the question was one of diagnosis. The friends had become impatient in consequence of the promises made by the physicians all having proved false. When I saw her I had no doubt, although I could make out nothing more than the physicians did, that it was a case of the kind I have been describing to you, and I had nothing to say but to recommend further expectation; and the case ended as the case I have been reading to you ended. The abscess at last pointed, as it usually does, in the groin, and the woman was very soon cured by Nature.

The case I have just read to you was a case of remote parametritis, but the parametritis was continuous with the uterus and fixed it. The case I am now to read to you is a case of remote parametritis without continuity. It is not a case of the ordinary kind of pelvic abscess; it is not a case of the kind that I first illustrated, inguinal parametritis; it is one of remote psoas abscess.

M. A. F., aged thirty-three, single, had a seven months' child seven weeks before admission to "Martha." Has never been well since. Catamenia appeared about five weeks after delivery. Bowels regular. Micturition natural. In evening, pulse 116; temperature 102.2°. Looks very ill, is worn and emaciated. Complains of pain in the right thigh shooting into the hip, and that she cannot walk. The right thigh is drawn up, being flexed about fifteen degrees. The upper half of the thigh is rounded and swollen to at least twice the size of the other. The swollen part is tender, but no special hardness can be found in it. The slightest touch beneath Poupart's ligament causes acute burning pain down the inside of the thigh. No defined hardness can be felt above Poupart's ligament, but there is great fulness there and extending upwards on the right side of the belly. The uterus is not fixed nor tender; neither is there any hardness nor tenderness around it. The right labium majus is swollen, being œdematous. She was put on a water-bed, ordered to be well fed, to have a morphia draught at bedtime, to have the lower abdomen constantly poulticed. Eight days after admission a pointing abscess was detected in the upper and anterior part of the right thigh. Next day it was opened with antiseptic precautions. About a pint and a half of fetid pus flowed. At the same time there was considerable hæmorrhage, apparently venous, and certainly not from the wound, which was made in thinned skin. The bleeding was arrested by pressure. Pressure in the region of the psoas and iliacus causing gushes of pus, made it plain where the abdominal part of the abscess was. Similar evidence showed that this part was in a few days completely evacuated and healed; but the large abscess in the thigh, whose extent was very ill-defined, required careful strapping and bandaging to secure its evacuation and healing. Now, six weeks after the abscess was opened, and fourteen weeks since her confinement, there is scarcely any discharge. The woman is rapidly regaining good looks and flesh.

Here is a case in which you have a large abscess forming in the right lumbar region, the region of the right psoas muscle; not a pelvic abscess. How the pus found its way down into the thigh we can only conjecture, as nothing abnormal was to be felt par vaginam. The pus probably advanced along the psoas muscle, and so got down into the region of the great internal muscles of the thigh. It did not pass through the pelvic cavity. This is an unusual route—a route into which it is forced, I believe, purely by mechanical circumstances; the pus burrowed because it found its easiest progress and accumulation by pressing downwards in this way. In this case there is also to be noticed the bleeding on opening the abscess. The bleeding made me regret that I did open it, because I think it probable we should have avoided this bleeding if we had let it alone; and in this case—I do not say in every case—I believe the opening would not have been delayed twenty-four hours. If the bleeding had been slight I would have thought nothing of it; but the bleeding was decidedly considerable, especially when you keep in mind the emaciated and exhausted condition of the woman. In this case I would remark to you what I have mentioned already, and what is described in the case—the evidence of an immense lumbar abscess; but no lumbar abscess could be felt on manipulating the abdomen. That, naturally, might lead to great mistakes were you not aware that it is not an uncommon condition, a large collection of matter present in this situation, but which cannot be made out by the examining prac-

titioner's hand.   You will remark in this case another peculiarity,—
that the pus was intensely fetid.   This putrefaction of the pus is
difficult to account for, for certainly there was no communication
with the bowel, or with any viscus, and yet putrefaction occurred.
This kind of occurrence forms a difficulty in connection with Lis-
terian antiseptics.   How did the pus putrefy without any route for
admission of germs?   The only explanation I can suggest is one
that I have heard, namely, that while healthy tissues will not allow
germs to permeate them, such morbid tissues as this woman had,
separating bowel from the cavity of the abscess, did allow germs to
pass.   That is, however, a hypothetical explanation.   No doubt the
fetid pus in this abscess led to great aggravation of the woman's
symptoms.   The putrid ichor was absorbed into the blood, probably
in considerable quantity.   This view is confirmed by an observation
which Mr. Garstang pointed out to me,—that on opening the fetid
abscess (the fetor rapidly disappeared, lasting about a day and a half)
improvement immediately followed, the temperature falling from
99.5° up to (in the evening) 102.5° down to natural, and that in a
few hours; and only once rising (ten days after the evacuation) as
high as 100.5°.   And this rising was due to some intercurrent affection which we did not discover.

Before concluding, I say a few words with regard to the treatment
of these inflammations and abscesses.   There is really very little to
be said.   The treatment of parametritis in all its forms is almost identical with the treatment of inflammation or an abscess in any other
situation—antiphlogistics; poultices; occasionally, in the phlegmonous
form, the use of blisters; and, lastly, the use of the knife.   It is only
on the use of this last that I shall here make a few remarks.   All
gynæcologists agree in discouraging resort to the knife in these cases.
I have often seen the knife used in the manner which, when we speak
of tapping, is called dry-tapping; the practitioner, not recognizing
the occurrence of parametric phlegmon, where there is no abscess,
and thinking he will hasten the progress of a case by driving his knife
into it.   But it is not the liability to mistakes of this kind that induces all gynæcologists to discourage opening parametric abscesses of
all kinds; it is their clinical history, which shows that in the great
majority of cases they are better let alone.   These remarks, however,
chiefly apply to pelvic abscesses, and there the danger of opening is
far greater than in remote parametric abscesses, such as I have been

going over to-day. The observation does not apply so rigidly to remote parametric abscesses, for it is frequently advisable to proceed to evacuate such in order to hasten the progress of a case that might otherwise linger for a very long time. I advise you, however, to be sure that you have an abscess to deal with, and to be sure that it is what is called "thoroughly matured" before you interfere with it.

VIII.

ON PAINFUL SITTING.

PAINFUL SITTING is the subject of this lecture. We have several cases illustrating it. In some of them painful walking accompanies the painful sitting; indeed, in the last disease that I shall mention to you, painful walking is more important than the painful sitting. Painful sitting is as good a name for a disease as dysmenorrhœa is, and quite as distinctive; but painful sitting is not a disease, nor is dysmenorrhœa; both are symptoms, and the term is used merely as an artificial arrangement of a variety of affections, just as dysmenorrhœa is so used. In both cases the designation is not a term of a pathological classification, but of what is called a nosological or artificial classification. The most common kinds of painful sitting are not to be considered to-day; only those that are observed particularly in women, and only those that are somewhat recondite. Such causes of painful sitting as an abscess of the vulva, an abscess of the perinæum, tender caruncle of the urethra, an inflamed gland of Cowper, are very common; and in them nobody requires to hear anything said about the painful sitting—that is a matter of course.

The first special cause of painful sitting that I have to consider is inflammation, not affecting the external organs, not affecting the vagina, but affecting the deepseated genital organs, the uterus and ovaries. This is not an infrequent cause; and the first case I am to read to you is a good example of it, an apposite example for us, because the poor woman came to the hospital declaring that she could not sit—that was her complaint; for her that was the disease. I shall read her case: "M. C., aged twenty-seven, admitted November 16th; married ten years; has had four children, the youngest two years old, two alive; the others died during teething. Has had no miscarriage. Catamenia began at thirteen years, and have generally been regular; nothing abnormal noticed about any of the later periods. About six weeks ago she was suddenly attacked with a severe pain in the right inguinal region, which has been present ever since when sitting. The pain is hardly felt at all when standing or lying down. This pain she ascribes to a kick on the belly; and I think it is a very probable explanation of it. About the middle of this term of six weeks she had a scanty thin brownish fetid discharge, which has since subsided, and is now imperceptible. It lasted for a week or a fortnight. The pain is identified by pressing on the perinæum, and subsequently by pressing the uterus digitally. The cervix uteri is nearly in its natural situation, patulous and hard, admitting the finger easily. It has an irregular hard internal surface. The uterus is fixed. The whole roof of the pelvis presents hardness, or dense fulness, which is tender." Now you can easily, from the record, make out that this woman has cancer of the neck of the womb. She knows nothing about that, and I believe does not suspect it; she thinks her disease was caused by the kick she got from her husband, and as for her the disease is painful sitting, I think she is quite right as to its cause. Somehow or other this kick was connected with an attack of perimetritis, an attack of inflammation around the womb, inflammation affecting the serous membrane, inflammation leading to the fixation of the uterus, which we found; and, so far as her disease consists in painful sitting, this inflammation is the cause of her disease. Attacks of inflammation, apart from violence, are quite common in connection with cancer of the womb. This woman's sufferings are caused by inflammation around the womb in the early stage of cancer of its neck. Now, I wish you to observe how clearly in this case the nature of the disease was made out. Firstly, a cancerous uterus is not a tender one. This woman's uterus was not tender where it was cancerous; it was the neighborhood of the uterus that was the seat of the tenderness, the seat of the inflammation. When the perinæum of this woman was pressed by the hand, while she was lying on her side, she at once recognized the pain of sitting. She felt the same pain as when the perinæum was pressed upon by the seat. She had not the pain when she lay down or when she was standing. Following up the pain, the finger was introduced into the vagina, and found the same pain was produced by pressing upon the inflamed and tender parts near the womb. There could thus be no doubt of the nature of the disease. Of this part of her disease, which, unfortunately for the woman, is not the major part, she will get rid by suitable treatment, especially by continued lying in bed. She is, indeed, already nearly well.

In connection with this case I shall state the theory of this painful sitting; and a very easy experiment explains it. It is not generally recognized that the bowels are pressed upon by sitting; but it is a fact, as this case illustrates. When a woman sits upon a seat, the pressure upon her hips, even although the deeper parts are protected by the tuberosities of the ischia, communicates pressure to the deepest parts in the pelvis; and, if those parts are tender, pain is the result. The experiment that I alluded to, as demonstrating what I have just said, is to place the hand upon the hypogastrium while the perinæum is exposed. If you press with it in the direction of the axis of the brim, you push down the perinæum and the hips. A very slight pressure upon the hypogastrium makes the perinæum bulge, makes the hips descend. Of course, when the hips are pressed upwards, or the perinæum is pressed upwards, you have an influence which is, in like manner, communicated back to the hypogastrium; and thus you have pain if the parts are inflamed. This is illustrated in many cases of ovaritis and metritis of all kinds. This part of painful sitting is a separate thing from the injurious influence of sitting. That I am not speaking about. The injurious influence of sitting is a subject I may illustrate at some other lecture. What I am speaking of now is painful sitting, and the injurious influence of sitting is a much wider subject than what I am now considering.

I go on, now, to another set of diseases, connected with the coccyx, which diseases are not peculiar to women, but are, I believe, much more frequent in women than in men; and they have got a collective name, which is also an artificial, not a pathological name—Coccygodynia. Now, the pathology of this department of painful sitting is so far advanced that I recommend you to give up the use of this name except as a proper word to express pain in the coccyx, for which no further explanation can be given. That is to say it is a neuralgia; perhaps not a pure or simple neuralgia, but yet a neuralgia; and a neuralgia is in the majority of cases a disease of which no further explanation can be given. Indeed, many of the cases usually included under coccygodynia are not diseases of the coccyx at all.

Occasionally the coccyx is the seat of inflammation; or its periosteum gets inflamed, and you have abscess around it. Of that disease I have not seen an example, but I have seen enough to show me that such a disease may exist. I have, for instance, seen a periosteal abscess extending from the point of the coccyx to the base of the sacrum, the whole length, which shows that such a disease as inflammation and abscess of the coccyx may occur. There is no doubt, indeed, that it has occurred. But the commonest cases of neuralgia of the coccyx or of true coccygodynia, although they have tenderness, or rather sensitiveness, as a symptom, have no inflammation, no inflammatory tenderness. Now this disease is common, and it is common in men as well as in women. I have seen cases of it in men, although I come very little in contact with that sex. In men it generally arises from constipation or some disorder of the rectum, such as hæmorrhoidal congestion. I may mention an example of it as it occurs in women. A young lady in her first pregnancy, enjoying perfect health, was sent to me, only two days ago, by her husband, because she could not sit. When she came into my room she laughingly said it was an absurd complaint, but she could not sit. It was easy to make out that she had this tenderness,—not inflammatory tenderness,—this sensitiveness, rather, of the coccyx. Now, this disease is generally easily cured, or rather it goes away, and the treatment of it is scarcely worth describing. It is the use of laxatives, hot bathing, sedative applications. In a severe and persistent case you may try the hypodermic injection of morphia, and it has been said to cure the disease. Whether it has done so or not I shall not answer for; perhaps time would just have cured it equally efficiently. The disease is essentially a come-and-go one, and it is very difficult to judge in such diseases what is to be attributed to treatment and what to time. These are the commonest cases. Other cases, however, are not rare; they arise chiefly from injury, and they seem to affect the sacro-coccygeal joint, and still more its ligaments, and the sacro-sciatic ligaments especially. Of this affection I shall give you an excellent example. Mrs. L., aged thirty-two, married for two years, had her first child nine months ago. During the second stage of labor she had intense suffering, especially during pains, in the region of the coccyx, where she has still all that she complains of. Ever since her confinement she has had the pain very severely during defecation, but now it is less than it was at first. When she began to get up, sitting brought on the pain, and she had to give it up entirely; but lately the pain in sitting has diminished, and now it is entirely gone. On examination the coccygeal region is easily identified as the seat of all the pain. There is no swelling nor dislocation of the bone. Pressing on it increasing flexion, as in sitting, causes now no pain; it did so at first when the parts were more tender; but extension, so as to bring the least tightness of the sacro-sciatic ligaments, brings on the

6

well-known pain. Pressure on the ligaments themselves, to tighten them, also induces the pain.

This is a very clear case, and the disease is gradually disappearing. The only pain that remains is produced by stretching the sacro-sciatic ligaments. I have no doubt that the disease in this woman is some sort of inflammatory rheumatic condition of these ligaments. Neither have I any doubt that she will get quite well; she is in the process of getting well. You observe that, in this case, the disease was brought on by injury sustained during labor. This intelligent woman's account of the second stage of her labor, and of the pain in the coccyx, leaves no room for doubt that then the disease was produced. The pain she suffers now is the same she suffered then, only much less. This case I recommended to be treated by hot bathing, by keeping the bowels easy, so that large masses of fæces might not descend and cause great extension of the coccyx. No more treatment was demanded, because the case was progressing slowly in a very satisfactory manner. Should it prove obstinate, I would be inclined to recommend to this woman to have the sacro-sciatic ligaments divided at their attachments to the coccyx—an experiment which is well worth trying. It has frequently failed to cure this disease; but then its failure may be because the treatment was used in cases for which it was not appropriate. This disease is only in that condition of progress in which we are differentiating the various kinds of it. You are not therefore to condemn this treatment altogether, on account of its failures, till the disease is much better known and the proper cases for this operation of dividing the ligaments are made out; if there are any proper cases. Its success in some cases surely indicates that there are proper cases. I should be inclined, in the case of this lady, to recommend its trial if the disease proves inveterate.

Before I pass further on, I shall make a statement to show you how imperfect yet is our knowledge of this disease. An eminent author, calling this disease coccygodynia, which indicates the want of recognition of the various diseases included under that name, says that a characteristic of it is that, while pressing upwards or from the outside is painful, pressing downwards or from the inside producing extension is not painful. The opposite was the condition of the patient in the case I have been describing to you, and the opposite is the condition I would write down if I were making such a general statement. I would rather be inclined to say that you had always pain from pressure extending the coccyx, pressing from within; only occasionally pain from pressure pushing the coccyx upwards, or flexing it by pressure from without, as was true of the earlier part of the history of our case.

I come now to another disease, of which we happened to have two examples in "Martha" almost at the same time—indeed, I think they were the same day,—dislocation of the coccyx. In these dislocations you have no pain, you have no tenderness; you have merely inconvenience which amounts to pain, inconvenience arising from the abnormal position of the coccyx, and which you will see admirably illustrated in the dislocation forwards which I am to read to you presently. It would surely be a great mistake to call this disease coccygodynia. When a man has a dislocated arm you do not call it humerodynia; neither should you call this coccygodynia; it is dislocation of the coccyx. The first case, E. G., aged thirty-seven, was admitted into "Martha" for carcinoma uteri. She made no complaint of her coccygeal region till her attention was called to it. Then she described herself as aware of something wrong there; and this has troubled her only since her last confinement, when she was delivered by instruments at the end of the seventh month of pregnancy. The coccyx is dislocated backwards, and is in a state of great unnatural flexion; it is only slightly mobile. This case, you see, is, like the last, a traumatic case; but it is also very unlike the last, for in this case you have no kind of inflammation; you have merely a dislocated coccyx. Dislocation is recognized by feeling externally the base of the coccyx, by passing the finger into the rectum, and feeling internally the point of the sacrum. Two parts which ought to be in contact are separate from one another, and the dislocation is backwards. In this dislocation backwards the coccyx is flexed. In this case nothing was done. It is recommended by some authors to reduce the dislocation. But it is another thing to do it. Indeed, it cannot be done. At the time of the accident, possibly, it could have easily been done; and now the attempt would be vain. The hold you can get of the part is so slight that you can exert no adequate pressure to tear up all the connections that are now formed in the situation of the sacro-coccygeal joint. Reduction, I believe, if it is ever to be successful, must be done either after dividing the connections between the sacrum and coccyx, or at the very time of the accident.

The next case I have to give you is a case of dislocation forwards, and you will see that this is a much more important accident. The

dislocation forwards in the case I am to read was traumatic, as in the
last case.   Mrs. N., aged thirty-eight, married for three years and a
half.   Has had three children.   During her last confinement she
required some extraordinary assistance, which she could not describe
to make intelligible.   Ever since that time she has had pain in what
she calls her "tail."   The pain is now almost gone, and she would
say nothing of it were it not that sitting brings it on ; and she wishes
to have her power of sitting restored.   On examination, the coccyx
is found in a position of extension, pointing downwards and projecting against the skin ; it is not tender.   Further examination finds its
motion very restricted, and it is dislocated forwards.   You can easily
understand that dislocation backwards with flexion is a comparatively
innocent matter ; but if you have dislocation forwards, and the coccyx
pointing down upon the perinæum, as it did in this lady, you can readily
understand how very soon aching would come on after sitting upon
the point of the coccyx stuck into the seat, if she sat otherwise than
upon a single ischium.   All patients suffering from this disease or
any of the allied diseases sit in a peculiar manner upon the edge of
the chair, resting upon the ischium next the chair.   In this way they
escape the pressure upon the perinæum.   Surely it is extremely de-
sirable that this very great disability should be cured.   In this case
I made no recommendation but one.   The only thing that could cure
this woman was to remove the bone, or at least to set it free, to put
it into some other position. I should think the simplest matter would
be to take it away altogether.   In a case like this, did the woman
not get accustomed to the state in which she is, I should certainly
have no hesitation in dividing the connections of the sacrum and
coccyx, putting the coccyx into a more convenient position, or removing it altogether.   Removing the coccyx altogether has been
recommended for coccygodynia, and I am not to advise you never
to resort to it.   As little do I advise you to resort to it ; because, so
far as I know of it—and I know a good many cases—it has proved an
extremely unsatisfactory proceeding.   In the case of this lady, I have
no doubt it would be satisfactory, and cure her with very little risk,
almost none.   As I have already said, the various conditions of
painful sitting are not sufficiently recognized so as to enable us to say
that coccygodynia is to be treated in any, as yet undefined, class of
cases by excision of the coccyx.   This is not a case of coccygodynia ;
it is a case of dislocation of the coccyx, with manifest easily accounted

for painful sitting, and with a manifest cure to come from cutting out the offending bone.

I now come to do little more than mention a very interesting case of fracture of the sacrum and dislocation of the lower part, which made sitting impossible, and produced difficult walking for a long time. M. B., aged forty-seven. Eight months ago she fell from a ladder some twelve feet on her sacrum. She was confined to bed in consequence for a fortnight, having been picked up senseless. After this she was able to walk, but not so well as formerly. She has had difficulty and pain in defecation ever since. Complains of pain in her sacrum. The upper bones of the sacrum are normal, but at the junction between the third and fourth there is a sharp angle—a little more than a right angle—formed by the unnatural projection forwards of the lower two bones of the sacrum and the coccyx, which latter is itself movable. The uterus is in normal position and direction, but with mobility much impaired. The right side of the pelvis is natural, but on the left and behind is a dense hardness, rounded posteriorly, nodulated in front towards left side. Per rectum, the angle of the sacrum can be distinctly felt, the fourth sacral vertebra being dislocated forwards, the dislocated portion being directed to the left. The rectum runs to the right of this part of the sacrum, and the induration above mentioned apparently starts from the left side of the sacrum, though part of it is in front of the rectum, between it and the vagina. This is a rare accident, the only one I ever saw of the kind. Were it not recognized, the physician examining the internal organs might be led to form very erroneous ideas as to the nature of the woman's disease; but there can be no difficulty, when we recognize the fracture and dislocation of the sacrum in ascribing the morbid conditions internally to the fracture and bad healing of the bone. Is it evident that the fracture led to some effusion on the left side of the uterus, and some adhesions in Douglas's space causing hardness. This accident, were the woman young, might produce very great difficulty in delivery, and it would require very careful consideration if a woman having it contemplated marriage; still more careful consideration if she were in the family-way. On that subject I have not time to enter.

The next case I have to mention to you is a still rarer one,—not fracture, but dislocation of the spine upon the sacrum,—a case of spondylolisthesis. This word does not describe the region of the affection, but its use is confined to the conditions I am to describe.

There is no fracture here, for you have a joint, and the bones slip upon one another. The first bone of the sacrum and the last lumbar vertebra can be mutually dislocated without much fracture. There might have been fracture, but we found no evidence of it.

E. H., aged sixty-three, admitted for carcinoma uteri. She had a considerable time ago a fall from a trap-door on her back, some fourteen feet. She was stunned, and afterwards could not walk for many days. Ever since she has had pains all over both legs, but no loss of sensation or motion. At the first bone of the sacrum is a prominence, continued downwards into a strong sacral convexity retiring within the fold of the buttocks, the sacrum being unnaturally curved forwards as a whole. The lumbar spine is in a state of slight lordosis. Nothing additional is made out per vaginam. The conditions indicate spondylolisthesis, or dislocation forwards of the spine upon the sacrum.

Before concluding this lecture, I have a few words to say upon a condition of the joints of the pelvis, which is rare as a disease, and which interferes with sitting and walking, especially the latter; that is, relaxation of the great essential or intrinsic joints of the pelvis, the symphysis pubis and the two sacro-iliac joints. These joints in the end of pregnancy become naturally juicy and loose, and a considerable increase of motion is permitted in them. The loosening of these joints becomes morbid very rarely. When morbid it has been found sometimes, in a few recorded cases, to be so extreme as to produce hopeless lameness. The joints have been so relaxed, and present such an amount of mobility, that by no contrivance can they be fixed so as to enable the woman to stand. Cases of that kind are among the greatest of rarities, but cases of slight yet extraordinary loosening are not very rare. They are recognized or diagnosed with great difficulty. You are led to suspect the existence of the condition by finding that the disease dates from pregnancy; it may be not from the first pregnancy. The last case which I saw was a case beginning in the second pregnancy. The next thing that leads you to suspect the disease is to find pain complained of in the symphysis pubis, or in what the patient calls "the bone" in the mons veneris, and in the two sacro-iliac joints, or in one of them. The pain of the symphysis pubis and in one of the sacro-iliac joints almost invariably go together, but both sacro-iliac joints are not invariably affected. What is the difficulty of recognizing this disease? There is no difficulty in a case of extreme relaxation. Then the woman can find the disease out for herself; but in a case of slight relaxation it is a matter of great nicety, and you have frequently to put the woman through a variety of evolutions before you can satisfy yourself that these joints are moving. I have found it generally vain to try to make this out in the symphysis pubis, partly on account of the disagreeableness of the proceedings. The proceedings are extremely indelicate; but only in a certain sense, for there is nothing truly indelicate that forms part of a duty; but they are extremely unpleasant, and the word "indelicate" implies a part of the unpleasantness. Besides, when I have attempted to diagnose the movement of the symphysis, I have been extremely ill-satisfied. In the case of the sacro-iliac joint there is no embarrassment; the difficulty is in being quite sure of the movement. In a healthy woman you can make out no movement. You start from that. If, then, we find distinct movement, we may be sure that there is this morbid condition. This distinct movement is to be ascertained by seizing the haunch-bone, and, while the spine is fixed, trying to make it move a little; or you fix the haunch-bone, and make the woman move her spine; and then you can see or feel, while the haunch-bone is fixed, a distinct movement of the spine upon it by making the woman change her position. I advise you not to be sure you make it out till you have perfectly satisfied yourself. Supposing you make it out, is there anything to be done? Like many others, this disease is fortunately frequently spontaneously cured. It is natural to expect that, as in a cow the moving knucklebones get fixed again after parturition, so in a woman the movable haunch-bones will get fixed after parturition; and the same may happen more slowly in the extraordinary or morbid cases I am describing. In such cases I have always encouraged a woman to walk, to brave out the pain if she could, because the irritation produced by the walking may conduce to the refixation of the joint. Cases of this kind do get better. The bones do get fixed again. Until they get fixed there is one means which is of great value, that is, a very firm bandage around the pelvis. You give an artificial fixedness. Now, you will find it very difficult to get a woman to wear this bandage, because it is extremely unpleasant in itself; and it is only after she has found the advantage of it that she will consent to wear it. The bandage must be made, not of ordinary bandage materials, but of horse-girth stuff. This is put round the pelvis, and strapped as tightly as the woman can endure; and if it is to be of any use it must be inconvenient, because, in order to be fixed upon the proper part, it must descend to a considerable extent upon her limbs,—that is to say, it must come down to, or even a little below, the trochanter major,—and this makes walking very disagreeable. I have seen cases in intelligent women where I can have no doubt of the real advantage of this bandage — where, indeed, the woman could not walk without it.

IX.

ACHING KIDNEY—PYONEPHROSIS—STRICTURE OF URETHRA.

THE first subject of this lecture is Aching Kidney. I shall read to you no individual case of this disease, because in the class of patients that come to St. Bartholomew's, it is not considered grievous enough to secure a bed in the hospital. Among the better classes, where diseases are often unjustly appraised, it is often regarded as of the greatest importance and interest. We have had many cases in "Martha" of aching kidney, but in them this affection has been merely an epiphenomenon, or a part of other diseased conditions.

This disease is sometimes, both in men and women, very easily recognized. There are achings in cases of what is called floating kidney. The patient can put her hand upon the lump, and say, "Here is the pain," and there is no difficulty in recognizing the disease. But there are some cases in which the disease is difficult to identify. In pregnancy, for instance, right or left hypochondriac pain is very frequent. In many cases I have been able to be quite sure, from the history before and after pregnancy, that the disease was not to be classified in the vague way that is implied in giving it the name of hypochondriac pain, but that it was really aching kidney. In pregnancy you have opposite conditions to those in floating kidney in ordinary circumstances, for if pregnancy is advanced, you cannot get at the kidney to feel it and identify its position. Here I may remark that, while the disease often occurs in pregnancy, yet some women who are liable to it do not suffer while in that condition. A patient, now under my care, has tender aching right kidney, which began fourteen days after last confinement. Before her present pregnancy began she had had two attacks of it; but during pregnancy she enjoyed perfect health.

The disease in women is not a rare one, and its characters are the following: One or other kidney is the seat of pain. It is not a

neuralgic pain ; it is a heavy, wearying pain, deep in the side.   It is
in the region of the kidney, and in many cases (as I shall presently
tell you) you can easily identify it as being in the kidney itself.   It
is not generally that kidney pain which is a familiar symptom of calculus.   In such cases the pain is the pain of the pelvis of the kidney.
You have in the region of the small ribs posteriorly a boring, or a
nail-like pain.   Patients with aching kidney generally point to the
hypochondriac region, not to the back, as they often do in cases of
calculus in the kidney.

This renal pain is frequently accompanied by pain in the corresponding lower limb, referred most frequently to the course of the sciatic
nerve, sometimes to the course of the anterior crural.   The pain is
often accompanied (and you will find this to prevail throughout all
the subjects of this lecture) by irritability—I do not say disease—of
the bladder ; and it is frequently accompanied by pain in the course
of the ureter corresponding to the kidney affected.

The renal pain is not rarely present only during the monthly
periods ; and when it is present only during the monthly periods it
may be classed with that disease, which is very ill-defined, called dysmenorrhoea.   It should never be placed there, unless you wish to use
the word dysmenorrhoea in a very wide sense.   If we use the word
as including aching kidney, we might as well use it as including
headache—a use which would be in accordance with what is extensively done by writers.   This renal disease often eludes the examination of the physician, because it occurs, in many cases, only during
the monthly periods.   In all cases it is then aggravated.   I do not
think I have ever seen a case in which the patient did not volunteer
the statement that the pain was worse at the monthly time.

Now, you naturally ask, What has the kidney to do with the menstrual function ?   And upon this interesting subject I shall make a
few remarks before I go further.   Embryologically, the urinary and
the genital organs are closely connected.   That you all know, and I
have not time now to enter upon the embryology of the subject.   As
the genito-urinary organs become developed they get separated from
one another, and their close connection does not strike the student of
adult human anatomy, forgetting the anatomy of the embryo.   But
in the adult you occasionally find the proximity of the organs maintained.   The kidneys are sometimes found in the pelvis, and cases are
recorded where kidneys in the pelvis, maintaining their proximity to
the genital organs, have been the cause of difficulty in labor.   Now,

not only have these two organs an embryological or developmental
connection, but they have an intimate connection in pathology. Of
that connection the disease I am now speaking of is an example; and
I shall give you another example, merely mentioning it. A woman,
after abortion or delivery at the full time, has an attack of parametritis.
This parametritis extends; and a favorite extension, as everybody
knows, is along the cellular tissue in front of the psoas muscle and up
to the suet. Cases have been very carefully observed where there
could be no doubt that an abscess of the suet was the result of inflammation of the womb—following an operation—following an abortion
—following delivery at the full time. This is another pathological
connection between these parts, and I might give you more; for analogous inflammations are observed in the virgin. It is worth while to
add that I have not distinctly traced the reverse morbid influence, or
renal affections producing pain or disorder of the genital organs.

It is not usual to find both kidneys aching, and I guess—I can use
no stronger word—that the left kidney is more frequently the seat of
disease than the right one. You are not left in your diagnosis in all
cases merely to identification of the seat of the pain, although that
may be sufficient. Frequently in the region of the pain you can find
distinct fulness; that is a very important physical condition that I
have not time to explain to you. It can scarcely be made out in a
fat woman; but in many cases this condition of fulness over the
affected kidney is easily recognized. In addition, swelling of the kidney or of the suet, or of both, is not rarely to be made out. The
physical examination of the kidney is too much neglected. It is not
in floating kidney only that you can feel the organ. In many women
who are not nervous, yielding themselves freely to examination, and
who are not fat, you can feel the kidney with distinctness; and in cases
of this kind you can frequently make out, as I have said, that there is
a swelling of the kidney or of the suet, or of both. There is also
generally tenderness, sometimes great tenderness.

Now you can scarcely mistake this disease, in a good example, for
any other. The diseases with which you are liable to confound it are
pyelitis and calculus; and the diagnosis is to be made out mostly on
the following grounds: In pyelitis and in calculus the pain is more
incessant; and in pyelitis the disease may go on acutely with fever.
Both these characters may be absent—and, indeed, are generally absent—in the case of aching kidney. In pyelitis and in calculus you
generally have pus in the urine in greater or less quantity; it may

be very little.   In calculus you generally have, in the history of the
woman, blood in the urine ; and this is generally connected with some
violence in the way of exercise, such as riding in a rough cab, or
having a fall.   In the case also of aching kidney, exercise frequently
aggravates the disease.   You can easily understand that a woman
taking rough exercise, as in an ill-built cab, will feel an aching, tender
kidney irritated by the exercise.   But this is not a very well-marked
symptom.   Most women who have aching kidney do not complain
of exercise, although some do.   Aching kidney is a disease of much
less gravity, and more amenable to treatment than pyelitis or calculus.

The treatment is to be conducted on the general principles of the
therapeutics of neuralgia or slight hyperæmia ; and these two conditions are not so very remote from one another as may at first sight
appear.   A neuralgia sounds as if it were something quite different
from a hyperæmic condition ; but that has to be proved.   The remedies I have found of most service in simple cases of this kind are
tonic regimen and tonic medicines, especially iron in the form of the
tincture of the perchloride, combined with mild diuretics in small
quantity, and especially the common sweet spirits of nitre.

Before I leave this subject I have to make a statement about the
importance of this disease—a statement that gives it a greater importance in exceptional cases than it has in the general run.   At present
I have three cases under my care where aching kidney was easily
diagnosed.   In one of them there are occasional appearances in the
urine of a small quantity of albumen.   These occasional appearances
of albumen are discovered by being looked for ; and the looking for
is stimulated by the occurrence of general ill-defined illness, or bad
headache, or something of that kind, which leads you to inquire into
the condition of the urine in a woman who has aching kidney.   In
a few cases of aching kidney of this kind I have detected the repeated
occurrence of albumen in the urine.   It occurs in generally small
quantity, but quite distinctly, and it occurs either without any casts, or
with few.   I have under my care at present a sufferer from aching kidney who recently became pregnant, who went on in pregnancy for four
months, and then albumen appeared in her urine in small quantity.
It appeared without any aching in the kidney, and it was not in the
form or under the circumstances in which it causes very great alarm.
It was a repetition of what had occurred long before her pregnancy,
while her kidney ached severely.   The albumen disappeared from
the urine entirely in about ten days.   She was then supposed, and

supposed herself, to be doing remarkably well, when a miscarriage occurred. Just as the albumen had distinct connection with the aching kidney, so had the miscariage. The miscarriage was caused, no doubt, by the death of the fœtus; the death of the fœtus was caused by disease of the placenta; the disease of the placenta was connected with a morbid or watery condition of the blood, which was probably present in this delicate woman in an exaggerated state. The disease of the placenta showed itself by the production of extensive yellow patches. These yellow patches are decolorized cotyledons, which have been thrombosed previously and rendered useless to the fœtus. The cotyledons of this woman's placenta became thrombosed one after another, until the placenta was reduced in useful area to such an extent as to lead to the death of the fœtus; and this thrombosis of the placenta had, no doubt, an intimate connection with the morbid condition of the kidney that I have been mentioning.

This condition of the kidney, with persistent or intermittent albuminuria, occurring in pregnant women, is not a very rare condition, and you are to distinguish it from the acute nephritis of pregnant and lying-in women, which is so frequently, or almost invariably, part of the great disease called puerperal eclampsia. I am not asserting that this disease does not form a minor degree of the same disease. I make no assertion about that; but to confuse the considerable, though often temporary, albuminuria I am now describing with the acute nephritis that leads to eclampsia would be a great error. In the pregnant woman I have been speaking of, the disease was temporary. It no doubt partly conduced to the death of the fœtus and the miscarriage, but there was no other harm to the woman, nor, indeed, the slightest alarm from any cause.

That is all I have to say upon the subject of aching kidney, and it brings me to the subject of pyonephrosis. The patient whose case is now to be under consideration, so far as her kidney was concerned, might be truly said to be suffering from aching kidney. But to give such a mere nosological name to her disease would be an injurious proceeding, not doing justice to our intelligence. For her there was nothing but an aching kidney, but we could easily make out that the cause of the aching kidney was pyonephrosis. We do not call it hydronephrosis, because we found that the tumor, which I shall immediately describe, was full of pus. It did not fluctuate, but it presented the feeling of fluid all over. There could be little doubt that it was not hydronephrosis. Hydronephrosis generally presents

a lobulated solid tumor, with more or less extensive portions exhibiting
fluctuation; and when the tumor is tapped you get out a filthy fluid,
often with urinous constituents. Before I read the particulars of
this case to you I may premise that the patient came into "Martha"
not for nephrosis, not for aching kidney, but for irritable bladder.

A. S., aged nineteen, was in Martha Ward in October, 1876, with
vascular tumor of the urethra. It was removed, and she was discharged cured at the end of the month. In March, 1877, she was
again admitted, with recurrence of the growths around the urethra
and hymen. They were cauterized, and she was discharged relieved
in April. In May, 1878, she was again admitted, and this time
under my care. She states that a few days after leaving the hospital
her painful symptoms returned, and are now worse than ever. She
has pain and smarting on micturition, and has to pass water about
every two hours. Four months ago she noticed a lump in her right
side. It is gradually enlarging. Before she discovered the lump
she suffered pain in the part for two months. It is constant, and,
though never very severe, has occasional exacerbations. The mam-
mary areola is of about the area of a shilling, and the mammilla not
larger than in a male. In the right flank is a lobulated tumor, the
chief lobe or nodule of which is just below the umbilicus and two
inches to the right side. There is a feeling of fluid contents; good
resonance below the tumor, and a streak of imperfect resonance between it and the liver. Impulse can be most distinctly obtained
between the renal region behind and the front of the tumor. Around
the posterior two-thirds of the orifice of the urethra there is arranged,
in a moniliform manner, a series of five crimson flat ulcers of about
the size of coriander seeds. They are slightly raised, and have
irregular, starred edges. They are supersensitive. The bladder
measures five inches from the orifice of the urethra to the fundus,
and is natural in point of sensibility and elasticity. The cervix uteri
is very small; the probe passes into the cavity two inches and a half.
There is no hymeneal obstruction. On May 9th the upper and right
lump was tapped with a fine trocar and aspirator. Nothing came
out. On May 13th the urine, acid, had a specific gravity of 1012,
faint cloud of albumen, slight deposit of pus on standing. May
18th: Under chloroform, the most prominent part of the tumor was
tapped with the aspirator trocar, and three or four ounces of extremely thick, viscid, purulent fluid drawn off, somewhat like putty.

No pain or tenderness followed the operation, and she was discharged on June 27th.

I have no idea what is the cause in this young woman of the obstruction of the ureter. Her disease consists in dilatation of the pelvis of the kidney, of its tributaries, and of the kidney itself, by pus. The nature of the pus extracted, and the general condition of the patient, suggest that the disease is at present obsolete, if not retrograde. That is a very important point, indicating the advice we gave. Cases of this kind are very serious, being very dangerous; and I have little doubt that sooner or later this young woman will have to call again for advice, for the disease she has is generally fatal. It is treated by one or other of three proceedings. One is to let it alone—which is a very painful resolution for all parties; the practitioner standing aside in impotence for substantital relief. Another proceeding, which has proved extremely dangerous, is to open the tumor and let the contents run out; and this is done in a variety of ways. The proceeding which we contemplated in this case was of a different kind—the excision of the kidney bodily; but we not only did not press this operation on the woman, we recommended her to go away without it. The operation is an extremely dangerous one, and the case at the time was not urgent. As I have said, however, the case will some time or other in this woman's life prove urgent if it behaves as other cases of the kind do, and the question of operative interference will come again to be considered.

That is all I have to say about the pyonephrosis part of this case, but the case introduces us to very important practical questions. You will observe that this patient came into "Martha" complaining of irritable bladder. It was for her a secondary matter that she had a lump in the side which ached; it was for irritable bladder that she had been in the hospital twice previously, and now for the third time —for the first time under my care. Having resolved not to interfere with her kidney, it was our duty to consider what we could do for her main suffering, her irritable bladder. In order to decide what was to be done for her irritable bladder, we had before us one of the most difficult questions in practice—Where is the disease? This irritable bladder, was it the consequence of the pyonephrosis?—Was it the consequence of the condition of the orifice of the urethra that I have already described?—Was it the result of disease of the bladder itself? Upon the decision depends the line of treatment. In this case we excluded disease of the bladder by examining it, and find-

ing (as you observe is recorded in the history of the case) that the
bladder was natural, not tender, and elastic. A woman suffering
from a slight degree of cystitis would have her bladder somewhat
contracted, exquisitely tender to the touch of the sound, and hard or
inelastic. We had no hesitation in concluding that it was not the
bladder that was in fault. We had next to consider whether it was
the kidney or not. That the kidney was the cause of the irritable
bladder in this case was rendered very improbable by the fact that
the irritable bladder had existed before the disease of the kidney
existed, and had not been aggravated by the increase of the disease in the kidney. Lastly, we had no hesitation in concluding that
the irritable bladder was the result of the disease of the external
genital organs, affecting in her at one time the urethra, at another the
urethra and parts external to the hymen; and, when she came to us,
affecting the urethra in the peculiar way I have mentioned. Were
this woman married she would infallibly suffer from dyspareunia,
producing vaginismus. The peculiar disease, of which this is an
excellent example, is like lupus in its history, recurring after it is
healed or extirpated. Here we have it, not producing dyspareunia
and vaginismus, because the woman is not married, but producing
irritability of the bladder, one of its most common consequences.

Notice the interesting circumstance that the disease at first presented itself as a caruncle of the urethra. The nature of that
caruncle, or the fact that it was not a common caruncle at all, but a
slight kind of lupus, is shown by the history of it; the presence now
of little ulcers about the urethra, the absence of caruncular swelling,
the presence of little ulcers around the hymen; contrasted with a
mere caruncular swelling when she first came to the hospital.

We settle the question as to the cause of the irritable bladder
by a study of the time and the order of appearance of the various
phenomena which we use to help us in forming a judgment, by the
characters of the phenomena, and by their severity. Using these
methods of judgment will not enable you to solve the question in
many cases. In this case it does enable us to solve the question.
Long essays have been written by eminent men—especially by sur-
geons studying diseases of the bladder in the male—upon this very
point, whether this irritable bladder arises from disease of the bladder, or of the urethra, or of the kidney; and, having read them, I
must tell you I consider that nothing could be more unsatisfactory.
We get further in the case of women than it is possible to do in the

case of men in settling this very important practical question; and you will observe it depends upon this settlement how you are to direct treatment. Your treatment will be in wrong lines unless you form a good judgment on this point. Now, here is a case of this disease not producing dyspareunia and vaginismus, but producing irritable bladder. In the lower animals irritation of the orifice of the urethra produces contraction of the bladder. Here is an example; and no doubt can remain upon the subject, for when this woman was cured of the irritation of the orifice of the urethra she was cured of irritability of the bladder; and when the disease causing the irritation of the orifice of the urethra came back, the irritability of bladder came back.

To conclude this case, I must tell you that we did nothing by operation. The patient had been thoroughly, even heroically, treated twice; yet the disease came back, and we did not feel disposed to begin again. If she returns we may reconsider that, and give her another chance of getting rid of these irritable ulcers, this lupoid disease about her vulva.

Lastly, I come to a case of Stricture of the Urethra; and I am fortunate in having it to relate, because, if I were to make a case to illustrate what I have been saying, I could not get a better one. A woman came to us suffering from irritable bladder. She had to make her water frequently, sometimes every few minutes. It was not a case of hours, but of minutes, and she could not get good sleep. On examining this woman we found that there was no orifice of the urethra in the natural situation. She had no history of syphilis, of operation, or of injury; yet there was no orifice in the situation of the urethra. A little to the right side of the natural position of this orifice was a very slight redness. A little surgical probe pressed against this redness entered the bladder. The orifice of the urethra, then, was strictured. On examining the woman's bladder we found it not expanded. It was a large bladder, but not larger than you frequently see in healthy women; but we found the urethra expanded. The bladder cavity did not begin at the internal orifice of the urethra, but its expansion began at the external orifice. There was no urethral canal. The bladder was not inflamed in any degree; it was soft, not tender, and large, though not unnaturally large. Now, here is a very plain case. A little operation was performed with a bistoury, enlarging the external orifice of the urethra, so that bougies, number 15, 16, and then 18, were passed into the bladder. Within two days the

canal of the urethra had re-formed itself; and from the moment of
the operation the woman was cured. She slept that night, she had
no irritability of the bladder at all, and made no complaint. She
remains cured. Much might be said upon this case, as illustrating
very important subjects both in uterine and in vesical pathology.
To-day I have only time to show you how important it is as indicating that disease at the orifice of the urethra may produce irritable
bladder. It confirms in a very remarkable manner,—as remarkable
as if we had made an experiment for the purpose,—the statement I
have made, that an irritable bladder may be due to a disease affecting
the external orifice of the urethra or its neighborhood.

X.

ON IRRITABLE BLADDER.

THE subject of this lecture is Irritable Bladder. Upon this sub-ject, in a former lecture—that upon Aching Kidney—I made some remarks which I shall not repeat now. They related to an impor-tant part of the matter, and referred chiefly to the importance, in the diagnosis of cases of merely irritable bladder, of taking notice of the order in time of the appearance of the phenomena of the disease, and of the severity of the different symptoms.

"Irritable bladder" is rather an ill-chosen name, because every bladder is irritable; every bladder has a peculiar faculty of sympa-thizing with diseases in neighboring organs, and, indeed, in some re-mote organs; the influence of sympathy is observed even in the case of emotions. But every bladder is not in the state of disease called irritable bladder—a condition in which the bladder is not only irrita-ble but irritated, and that generally not by disease referable to itself. An irritated bladder may be so merely by sympathy, or reflex influ-ence; and I shall give you cases where there is no other possible explanation of the irritation of the bladder than that founded on sym-pathy or reflex influence. But some irritated bladders exhibit a cer-tain amount of catarrh, that is, of superficial inflammation; and this catarrh may be truly called a secondary disease, not a disease of the bladder primarily—a disease, therefore, to be studied in connection with the cases of merely irritated bladder to which I am chiefly directing your attention. You know well that inflammation may be excited by sympathy. That is well illustrated in some cases of inflam-mation of the eyes, disease in one eye inducing disease in the other; and that kind of induced disease has much the same history, much the same characters altogether, as the characters of the simply irri-tated bladder that I am going to describe.

Some cases of irritated bladder are, no doubt, explained by the irri-tation being conveyed, not through contiguity, not through any ner-

vous connection, but through the passage of morbid products from
one organ to another.   For instance, there is no doubt that in children irritation of the bladder is frequently a sign of gravel.   That is
a well-known cause of violent pain in children, accompanied by
intense irritation of the bladder, and it is frequently explained by the
passage of the cayenne-pepper-like grains of uric acid through the
bladder and the urethra.   In adults the same irritating character is
justly ascribed to other kinds of gravelly urine or phosphatic urine.
But I am inclined to think that urine not decomposing, yet containing
ordinary morbid elements in solution, will not irritate the bladder.
It is, in general, only when you have the urine decomposed or carrying with it solid matters that you have a bladder thus irritated.

The importance of this subject has been fully recognized.   No diseases are more urgent than diseases of the bladder, on account of the
great suffering and inconvenience which they cause.   If you do not
diagnose properly a disease of the bladder, if you mistake a merely
irritated bladder for a more real disease of the bladder, your whole
treatment will be misdirected, and you will be, so far as your ignorance is blamable, a bad adviser to your patient.

A bladder may be sympathetically irritated by diseases of the kidneys, of the ureters, of the urethra, of the external organs of generation, of the pelvic organs; and it is impossible to exclude some conditions of the bladder itself as a source of mere irritation.

Now, what are the indications of mere irritation of the bladder?
To the patient the great indication is the frequency of urination.
This is often accompanied by pain in urination, or strangury.   But
frequency of urination is not of itself a proof that a woman has irritable bladder.   For example, a hysterical woman, when she is under
the influence of that condition, urinates frequently because her bladder is frequently and rapidly filled.   And it is not an uncommon
thing for diabetes to be mistaken, for a time at least, for irritable
bladder.   I have seen this repeatedly happen in consequence of insufficient attention to the circumstance that frequent urination is not of
itself a sign that a woman has an irritable bladder.   You must take
into account the quantity of urine; and in the case of hysterical and
diabetic women, if the quantity were taken into account, the error
would quickly disappear, the explanation of the supposed irritability
being at once given.   On the other hand,—and this is a still more
important remark,—the passage, frequently, of a small quantity of

urine is not of itself a proof that a woman has an irritated bladder; especially it is not a proof that she has a bladder which resents moderate repletion. When a woman passes a small quantity of water frequently, regularly, you may be pretty sure either that her bladder is small and contracted, or that she has an irritated bladder—that is, one which resents too soon an ordinary amount of repletion. But you have no proof, in the fact that a woman frequently passes a small quantity of urine, that her bladder is not enormously capacious. Some of the commonest errors in obstetrics arise from neglecting this. For instance, a woman with retroversion of the gravid uterus not rarely complains of irritable bladder,—that is, of frequency of urination and painful urination,—and yet her bladder may all the time contain a very large quantity of water, only a part of which does she pass. The same thing is true of other conditions of the bladder, apart entirely from pregnancy—conditions of irritated bladder from permanent overdistension or too great size.

In a case of simply irritated bladder, of the kind easiest understood, you have healthy urine. But, studying a case of irritable bladder, you are frequently called upon to pay attention to the condition of the urine, to its contents, such as mucus. Urine containing a large quantity of mucus is, at least at first sight, supposed to be urine from a bladder which is not only irritated, but in a state of catarrh; and this suspicion is increased by the circumstance, if it is present, of the mucus being mixed with pus. Still further is the fear increased if the urine contains blood; and still further if the urine is alkaline. Now, all these circumstances—a large amount of mucus, some pus, some blood, and an alkaline condition of the urine—lead you to suspect or believe that the bladder is not merely irritated, but also diseased, and that not secondarily merely. You have to consider whether the bladder may not be sympathetically inflamed in a slight degree; and, secondly, whether these products discovered in the urine may not be derived from other sources than the bladder. Mucus is seldom derived in large quantities from the ureters and pelves of the kidneys, so that a large quantity of it is more distinctive; but it is impossible to say where the limit lies between the amount of mucus that may be secreted from the pelves and ureters of the kidneys and that which is distinctive of vesical catarrh. Lastly, in order to complete a diagnosis of a case of irritated bladder, you examine the bladder itself. I think this is the most important part of the means to be employed, and I shall make special reference to it at the close of the lecture.

I propose now to give you some examples of this disease, most of which have occurred in Martha Ward during the last few weeks. I begin with a case which is extremely clear. A young woman, a long time under treatment elsewhere for menstrual disorder, consulted me, and the only thing I could find wrong about her was an aching left kidney. She had, no doubt, had menstrual disorder, but I could find, by physical examination, nothing to account for it. When I saw her first her urine was quite healthy. Some months afterwards she returned to me, and then her complaint was of irritated bladder, and she gave the best description of her symptoms by saying that she had to get up several times during the night in order to make water. The examination of her bladder revealed a perfectly healthy condition. The urine was limpid, without deposit, of a natural specific gravity, and healthy in every respect. But, on more careful examination, it was found that her aching left kidney was worse than it had ever been, and that her urine contained albumen in considerable quantity. Here, then, you have a case about which there could be no hesitation. There was nothing in it to account for the most prominent symptom, the chief complaint of the patient, but the condition of the kidney. In my lecture on Aching Kidney I gave you another example of the same kind.

I now come to a case of greater difficulty. E. M., aged thirty-two, not married, was admitted into Martha Ward on account of supposed disease of the bladder. Urine acid, 1012, containing a small quantity of mucus and still less of pus. Complains of having to pass water ever half hour, and the process is accompanied by pain which she feels greatly in the private parts; she also complains of pain in the loins, and down the inside of the thighs. She was ordered a saline laxative, to keep her bed, and to have three times daily an ounce of decoction of Pareira, half a drachm of tincture of hyoscyamus, and twenty minims of sweet spirits of nitre. Under this treatment her symptoms were in a few days greatly diminished without any improvement of the urine; indeed, close observation discovered it frequently tinged with blood, and always containing albumen, pus, and mucus. The bladder was now examined physically, and no disease whatever was detected in it. The patient's chief or only sufferings latterly were from pain in the region of the kidneys. A consultation with Dr. Gee resulted in the opinion that the disease was not in the bladder, but in the kidneys or their pelves. Of the truth of this result I can have no doubt. You see that the treatment removed the woman's bladder symptoms, but left her renal symptoms. The quantity of mucus and pus was small, not what you would expect in a case of disease of the bladder; and the end of the case, after successful treatment of the vesical symptoms, being persistence of the renal symptoms, we had no doubt that we had an example of a bladder irritated by sympathy with disease of both kidneys or their pelves.

I now come to examples traceable to the external organs, and I recall to your minds the case of stricture at the external orifice of the urethra, where the complaint was frequency of urination, where there was no disease in the bladder discoverable on physical examination; and where, on removal of the stricture of the orifice of the urethra, a cure was instantaneously effected. I will give you another example where the disease arose from caruncle of the urethra—caruncle of an irritable kind.

S. S., aged fifty, has been married thirty years. Catamenia ceased a year ago after previous regularity. Complains of forcing pain in private parts, and of frequent micturition when she is not in bed. Urine acid, specific gravity 1020, no albumen, some phosphates. Attached to the posterior lip of the urethral orifice by a large pedicle is a small red caruncle of the size of half of a small split-pea. It is extremely tender. It was removed, under chloroform, by scissors. Next day hæmorrhage, to the extent of some ounces, was checked by ligature of a small vessel in the wound. She remained in "Martha" twelve days after the operation, and had no complaint whatever after the caruncle was removed.

This, again, is an example, as clear as possible, of irritated bladder cured by the removal of its cause, the cause not residing in the bladder, but in the external parts. In a former lecture I recorded a case where we had slight difficulty in diagnosing a case of irritated bladder while the woman had pyonephrosis, which of itself was enough to account for it; but she also had disease at the orifice of the urethra. Now, in that case you will remember we made out that it was the disease at the orifice of the urethra that was the cause of the woman's great and chief sufferings, namely, from irritability of bladder. When the disease of the urethra was removed she had no irritation of bladder, although the pyonephrosis existed. When the disease of the urethra returned she was unaware of its return, but her irritability of bladder returned, and that was her great complaint.

I might give you more examples of disease of the external genital organs causing irritated bladder; but I go on to say a few words as to the most difficult part of the subject. I refer to cases where the cause of the irritability of the bladder is in the pelvic cavity or in the bladder itself. It is well known that the bladder sympathizes with all sorts of diseases in the pelvis, and its sympathy is evidenced by irritation. In the case of inflammatory diseases, and in the case of some non-inflammatory diseases, we refer the irritation to congestion of the bladder, or inflammation in a slight degree communicated to it from neighboring inflammations. It is a very frequent thing to read of irritation of the bladder in these circumstances accounted for by pressure or by distortion; that is, change of shape and position. In regard to this, which is a very important matter in connection with the study of flexions or other minor displacements of the uterus, my mind is not quite made up, but I am strongly of opinion that no change of position, no distortion, no pressure of an ordinary kind, causes irritation of the bladder. You will find the bladder without a trace of irritation, yet having every possible shape and every kind of displacement; and I see no sufficient reason for referring (as is often done) irritation of the bladder to its change of shape, or pressure upon it, or displacement of it.

Cases where the bladder sympathizes with disease in its neighborhood are well known to all; but there is a class of cases where the irritation seems to be in the organ itself. These cases are characterized by the too great size of the organ. This can sometimes be traced to a distinct cause. Sometimes it is not to be accounted for. In cases of this kind the patient occasionally has, in addition to irritation of the bladder, incontinence of urine—that is to say, the incontinence comes on at times, and irritation at other times. When these cases are watched it is frequently observed that the woman can retain her water for a very long time, and sometimes it is found that there is air in the bladder. This air gets admission sometimes through the catheter, but I have seen it in such cases present where this explanation was not tenable, the air having got in through the urethra directly. When air gets into the bladder you can easily understand that you are very liable to have the urine decomposing, especially as it may be long retained; and this aggravates the case very much. The simplest case of this kind that I remember to mention is one of a young woman who was brought to me several years ago, in whom the obstacle to marriage was that she was in the habit of wetting her bed at night, and that during the day she had frequently to make water. The disease was distinctly traceable to her mother's bad habit of punishing the girl when she was a child for wetting her bed, and to overdistension of the bladder beginning then. The bladder was enormous. A vesical sound could be passed into it—I forgot the number of inches—but it could be felt at the umbilicus. There was no other disease present. The urine was healthy, and the case was cured by keeping the bladder empty. In order to keep it empty the bladder required more than to be evacuated by passing a catheter. A bladder may remain well filled while the urine has free exit through a catheter. In such a case, in order to insure that the evacuation is complete, you have to squeeze it out through the catheter, as the sportsman does with rabbits he has shot.

That is a simple case, and I have seen more than one of that kind, such as one that came under our notice in "Martha" not many days ago. She was an out-patient, so that I cannot tell what effect upon her the treatment by regularly emptying the bladder has had. Her case was evidently one of this kind. I will read part of it as given in the letter which was sent along with her: "Many years ago she was a patient of Dr. T., when she had uterine displacement with bladder symptoms. Five years ago she had an accident, and since that time the bladder symptoms have been worse. She complains of pain in the region of the bladder. At one time she cannot hold the urine, at another she cannot pass it, and at another she has to pass it very frequently. There is nothing abnormal in the urine, but occasionally it is alkaline." This woman was found to have a bladder of greatly exaggerated capacity, but otherwise healthy.

In these cases the bladder is sometimes not only large in capacity, but hypertrophied. Many are of extreme difficulty, and I pass on to another important point.

Irritation sometimes does not occur when you would most expect it, when even the bladder itself is diseased. Of this I shall give you several examples. You remember a case of pyonephrosis to which I have already referred. That pyonephrosis did not bring on irritability in the woman's bladder. In a case lately in "Martha," a urethral cyst did not bring on irritable bladder, but the treatment for the urethral cyst brought it on severely for a time.

Mrs. A. D., aged thirty-one, two years married, no children, came into the hospital, complaining of dysmenorrhœa, which has been gradually getting worse since marriage. She has a retroverted, bulky uterus, and a tender, inflamed left ovary. A tumor of the size of a boy's marble lies in the vagina, connected with the middle of the urethra by a large pedicle. It is cystic. There is no complaint of irritated bladder, nor is there frequent micturition. The cyst was opened by bistoury and evacuated of its viscid, glairy contents. Next day she complained of difficulty of micturition. Four days afterwards the cyst was reclosed. It was reopened freely, and cauterized with nitrate of silver. This increased greatly the irritability of the bladder for a time, but it soon disappeared; and when she was discharged she had no complaint of her bladder.

I may mention a still more extraordinary case illustrating the absence of irritability. A woman died, under my care—perimetric abscess and tubercular peritonitis. The perimetric abscess was a consequence of parturition, and it burst into the ileum. Simultaneously with the diminution of the abscess the urine became bloody, and carried with it a large amount of pus. I never doubted that the abscess had burst through the bladder. Pus was never observed in the stools. A post-mortem examination was made, and there was found no communication between the bladder and the abscess. The bladder was only slightly contracted, and the whole of its mucous membrane was dark red, in a state of the highest degree of catarrhal inflammation, secreting pus and also exuding blood. The woman had no irritability of bladder; she never complained of that organ.

Another case is that of a patient under my care, in hospital, with great hæmaturia. The case was diagnosed as being not one of disease of the bladder on account of the physical examination revealing a healthy condition, so far as it could be made out. This woman died suddenly, and her bladder was found to be everywhere dark red; and on its internal surface there were several nodules of soft cancer.

These cases show you the extreme difficulty of this subject, but they are so rare that they do not greatly diminish the confidence that you can place in the means of diagnosis that I have been describing. Before I pass on I shall tell you another curious condition in which a bladder ceases to be irritable. You know that, in cases of ulceration of the bladder, suffering is sometimes so intense that life is scarcely worth maintaining. I do not know any more dreadful picture of incessant agony than that of a woman suffering from chronic ulceration of the bladder of the kind that I am referring to. I remember well a case of this kind, in which the woman herself prevented me from opening her bladder to see if making a vesico-vaginal fistula would relieve her dreadful sufferings. I was not sure that it would have relieved her, but I hoped that it might. Years afterwards the woman got married. After her marriage a great ulcer broke out in her leg, and she went to the surgical part of the hospital and had her leg cut off. When she left the hospital she came to me to tell me that she had no trouble with her bladder now. Nothing could have astonished me more than this announcement. She told me also of her marriage. I asked her how she made water. She said she never made water. She had no vesico-vaginal fistula; her bladder was a mere dilatation of the urinary passage, through which the urine flowed without being arrested in it. She never made water; she had stillicidium urinæ from the urethra; her bladder was a non-existent organ for her—it was extremely small. In that case the examination of the bladder was necessary in order to diagnose it.

I come now, lastly, to describe what I consider by far the most important point in the diagnosis of merely irritated bladder—viz., the physical examination of the bladder, to ascertain its healthy condition or the reverse. This is done by the use of a sound. I here show you a common vesical sound which I use for the purpose. By this instrument you ascertain the size of the bladder, its hardness, and its tenderness. I shall take the last condition first. In a healthy bladder there is no tenderness. You examine carefully, without rudeness, a healthy bladder; the woman is not aware of your doing so. Between this and the intensest agony you have all variations of painfulness. I know nothing more severe than the pain of examination of the bladder when it is even slightly inflamed. As a first result of your examination, you ascertain the degree of tenderness, or its entire absence, by the sound.

The next thing you do is to ascertain its softness or hardness. A healthy bladder has a considerable elasticity, so that when you touch the fundus you can push the instrument one inch, at least, farther into the bladder, and it is pushed out again by the elasticity of the organ. When a bladder is irritated it may retain this condition—it generally retains it when it is merely irritated; but when it is inflamed in the slightest degree it soon becomes hard, and it resists the push of the sound. In some rare nervous women examination leads to complete temporary contraction of the bladder, so as to be completely closed, resisting in this way the introduction of the instrument.

This is found as a persisting condition in cases of the greatest inflammation, as in the acute stage of gonorrhœal cystitis.

Lastly, you ascertain the size of the bladder. In order to do this you need not attach any importance to whether the patient has made water recently or not. The bladder does not contract to empty itself. The main use of the contractions of the bladder is to announce that it is time to empty it, to call the woman to the bedside. If then you pass a sound into a healthy bladder to measure it, you must measure from the external orifice of the urethra, because you do not know exactly where the internal one is ; and in a healthy woman the sound is easily passed about five inches. In a case of chronic cystitis a very common measurement is four inches. In a case of acute gonorrhœal cystitis, with strangury, you very likely cannot get the instrument into the bladder at all, or, if you do, you will only have a measurement of two or two and a half inches.

To conclude, you can easily see that if a case comes before you as cystitis, and you find that the bladder is healthy, that it is large enough,—not too large,—that it is not tender, and that it is elastic, you have in these circumstances almost certain evidence that the woman's bladder is healthy, and that her great symptom, which may naturally give the nosological name to the disease, irritable bladder, is a mere symptom, and not the essence of the disease.

XI.

ON VAGINISMUS.

WE have had in Martha Ward recently several cases of vaginismus, and a case of secondary vaginismus forms the text of this lecture.

What is vaginismus? It is one of the numerous diseases that occur in two forms, either primary or secondary. When the disease is primary it is a pure neurosis; that is, we can find nothing visible or tangible to account for it. When it is secondary it is not a pure neurosis; it is a neurosis, but it is a neurosis for which we can in some degree account. This vaginismus is a neurosis of motility, and it consists of spasm. It may be called spasm of the vagina, for that is the part that is affected or changed. The spasm of vaginismus is, so far as it affects the voluntary muscles, a tonic spasm. The voluntary muscles that it affects are the constrictor vaginæ and the anterior part, if not the whole, of the levator ani. One result of the spasm of these muscles is complete closure of the vagina as a passage. This tonic spasm of the voluntary muscles has generally been regarded as the whole of the spasmodic part of the disease; but the affection in a bad case is so severe that I am inclined to think there may be other spasms of involuntary muscles concurring to produce the condition of a woman suffering from vaginismus, which I shall immediately describe to you. In the diseases of women there are many spasms of involuntary muscle; the most violent spasms producing the torture of extreme dysmenorrhœa being well known. It is also known that irritations which produce spasms of involuntary muscles in certain of the lower animals are identical with the irritations which produce the spasms that I am referring to—spasms such as I believe occur in vaginismus. For instance, experimental physiology has shown that irritation of the clitoris produces contractions of the uterine horns; and it is ascertained that irritation of the urethral orifice produces contractions of the fundus of the bladder.

It is, therefore, surely not going too far to suppose that, in the condition of a woman suffering from vaginismus, you have not only spasm of the voluntary muscles, the constrictor vaginæ and the levator ani, but also a painful spasm of the involuntary muscular fibres of the uterus proper.

When a woman is suffering from vaginismus, in a characteristic bad case, pain is produced by touching any of the external parts of generation near the vaginal orifice. The further touching of these parts throws the woman into a paroxysm of agony in which the spasms I have described occur. If the irritation is continued there results a state of opisthotonos. The woman is almost, if not altogether, insensible, and her recovery from the condition takes a long time; it may take hours to get over the disorder into which she has been plunged by the irritation that produces the complicated condition called vaginismus in an extreme case. The worst cases are simple uncomplicated cases where the disease is, as I have said, a pure neurosis of motility.

I have seen a case of simple vaginismus, not very severe, where the pain and subsequent aching were felt on one side only. On examination, digitally, the spasm could be felt to affect the left side alone; and it was pressure on the left side that induced this contraction of the anterior portion of the left levator ani.

There are other spasms in these parts which I shall not have occasion to lecture upon here, but which are so closely allied that I must mention them. There are a number of well-authenticated cases of spasm of the levator ani during sexual connection. This is not ordinary vaginismus, but it illustrates the subject. It is a painful spasm of the levator ani during sexual connection, in some cases producing incarceration of the penis. There are other cases of the same spasm (which I shall describe a little further on) induced by the process of parturition, and obstructing it.

I have given you a description of simple vaginismus, and you can easily understand from what I have said that, except in extraordinary circumstances, it is not discovered until sexual connection is attempted; it is therefore a disease which is most frequently discovered on marriage, when sexual intercourse is found to be painful and difficult, or impossible. If you consider the importance of this conjugal relation you can easily understand that, in a certain important sense, there is no more serious disease than this. It is a disease which involves no danger to life. The disease, if sexual connection is not attempted, is as good as absent; but in the case of married women it is a disease which is exceedingly important, apart from any influence it may exert upon general health.

This condition of the sexual relations is called dyspareunia—painful or difficult sexual connection. All cases of vaginismus are cases of dyspareunia; all cases of dyspareuna are not cases of vaginismus. You can easily understand that there are many cases of pain and difficulty in sexual coitus which are not vaginismus. All cases of vaginismus are proved by the dyspareunia; it is the dyspareunia that reveals the condition, or that leads to the investigation which discovers the condition.

I must say a little more about uncomplicated vaginismus. I have already given you a sketch of the disease; but there is a little more known about it, and before I dismiss it I must tell it to you. In a case of simple pure vaginismus I am not aware that you can discover anything in the temperament or condition of the woman in any way to lead you to expect its existence. The parts, when they are examined, are found to be in perfect health, perfectly well formed. In order to examine them, the patient must be put under the profoundest influence of an anæsthetic. In making the examination you will, in the great majority of cases, discover that the disease is not simple, but secondary,—that is, you will find something that more or less completely accounts for the disease. It is almost invariably accompanied by a diminution or absence of sexual desire; indeed, it is frequently accompanied by a negative condition of the sexual appetite—sexual repugnance. Simple vaginismus may come and go. Its coming on appears to be connected with the disappearance of sexual appetite, either as cause or effect. Such women, however, may conceive. It is a well-known fact that it is not necessary for conception either that a woman should have sexual desire, or that her vagina should be penetrated. The result of pregnancy illustrates the inveterate nature of the disease. One of the early cases on record, published more than half a century ago, gives an accurate account of the malady. It was a case in which the patient conceived, and had a child at the full time, and was none the better in consequence of parturition. I am myself aware of several cases of this kind; and this physiological or pathological fact has a very clear bearing upon the subject of treatment. In a case of simple vaginismus there is no cure, nothing of the kind, as the result of the birth of a child. Occasionally, in consequence of the great distension and laceration of the vaginal and vulvar orifices, there is a less intensity of the disease; but that is all.

It has been alleged that a woman suffering from this disease is liable to the same spasms of the voluntary muscles during parturition, and there are cases recorded where the parturition has been so difficult in consequence of this spasm as to require craniotomy. This kind of spasm occurs during parturition in women who have not suffered from vaginismus; but it is alleged to be a condition that is to be expected in cases previously affected by vaginismus. I am satisfied, however, that there is no good ground in actual observations for this expectation. In three cases that have come under my own care or notice very lately I have seen no such result. Perhaps the modern difference produced by the use of chloroform in painful labors may account for this absence of spasm during parturition under the influence of the anæsthetic, whose value in painful labor has been known for little more than thirty years.

The case of vaginismus that is the subject of lecture to-day is not a case of primary or simple vaginismus; it is a case of a much more common kind, a case of secondary vaginismus. In these generally slighter cases you can, by a careful physical examination, discover disease. The disease that occurs most frequently in newly married women is a painful red spot at the fourchette, occasionally also a fissure there. The red spot is at the anterior margin of the perinæum; the fissure may be either in the same place or in the fossa navicularis, or in the external or internal margin of the hymen. When this redness or fissure is touched, the woman can identify it as the source of her disease; she may say, "That is the part," and, on looking at it, you find the condition I have described.

The next most frequent condition observed, especially in newly married women, is vaginitis, either acute or chronic; and this, of course, accounts for the vaginismus without any difficulty. There is frequently, however, and especially in women who have been some time married, a chronic vaginitis which is in many, not in all cases so affected, the cause of the disease. Cases of severe, though not acute, vaginitis without vaginismus are not uncommon, and there is a case in "Martha" now. There are several other more remote causes which I shall not mention, such as the sensitive caruncle of the urethra.

The case I am about to read to you is an example of a kind of disease that is very far from uncommon, and which, I am sure, has escaped notice in many cases held to be examples of simple vaginismus, but which were really secondary. The disease is in outward appearance very slight, and requires thorough investigation to discover it. It consists in the presence of one or more little ulcerations, which appear to be healthy. They are generally situated round the orifice of the vagina beyond the hymen. Under treatment, or without treatment, they heal and break out in other parts. They are frequently accompanied by little hypertrophies—hypertrophies of bits of the hymen, hypertropies of the orifice of the urethra. They are intensely tender and sensitive; and, in order to their examination, the deep influence of an anæsthetic is necessary. What is the nature of this disease (which I do not think has been accurately described)? I am at a loss to say. Whether it is allied to eczema or to lupus I cannot decide. I think it is allied to lupus, and the characters that lead me to think so are these: first, the situation of the disease; secondly, the way in which it heals up and breaks out again; and, thirdly, the occurrence of these little nodular hypertrophies of the hymen, urethra, and other parts.

You are not to suppose that every woman with this disease has vaginismus; the association is not necessary by any means. A woman with a slight, or even a severe, degree of vaginitis may not have vaginismus. It is only when the pain and sensitiveness are extreme, or at least elicit the spasms, that the disease produces vaginismus. You will find many women with these and other ulcerations—some of them certainly lupus, others not—who have no vaginismus at all, indeed little or no tenderness of the affected parts. This is a very important distinction. No doubt it points to some important textural difference, which I cannot tell you of because I do not know it. Evidently there is great variation in these diseases, but I know of no difference in the general history or in the appearances on examination, except the sensitiveness and consequent production of a reflex vaginismus. Of this secondary disease the case that I have to bring before you is an excellent example. I shall not read it till the end of the lecture.

The last thing I have to enter upon is the important matter of treatment. In the simple, pure neurotic cases I am bound to say I know of no treatment that is of decided use. If the case is a slight one, the dyspareunia may be modified by an enlargement or distension of the vaginal orifice, but only slightly modified. Such distension can be easily effected by the surgeon. In a severe case any operation with this view is followed by no benefit. This is what I referred to when I spoke of the evidence in regard to treatment de-

rivable from childbirth. There, surely, you have abundant enlargement and laceration, tearing open of the orifice of the vagina and
vulva; and in a bad case of this kind there is no absence of the disease when sexual relations are resumed. In a slight case there may
be some improvement, and I have known diminution of the pain and
suffering follow the bearing of a child. There are, however, operations which I do not think have been sufficiently tried, and which
are justifiable, considering the desperate circumstances of a woman
suffering from intense vaginismus. I think it would be legitimate
to try the operation introduced long ago into practice,—the cutting of
the pudic nerve. I have seen the operation performed with no
benefit.

This individual operation was, in important respects, an unsatisfactory one, and did not contribute to settling anything; but I must
add that our knowledge of the therapeutic results of the division of
nerves is not, in this matter, very encouraging; and it would be no
easy matter to remove a long portion, say an inch of the nerve, to
obviate the failure of this operation from reunion of the separated
ends of the nerve-trunk. It has been proposed to remove the most
sensitive parts; I regard this proceeding, meantime, with no favor.
Operations of this kind have been frequently performed, and declared
to be successful. At present I have no doubt that the observations
were misinterpreted. It is quite easy to cure many cases of this disease. I have no belief in the cutting away of the hymen, or that
such operations have any influence in a simple, pure neurotic vaginismus. In secondary cases you are very hopeful in your treatment,
and your hopefulness is in proportion to the curability of the discovered tangible disease. In the great majority of instances occurring
immediately or soon after marriage, where you have the red spot or
the fissure that I have described, time alone, with rest of the parts, is
all that is required for their cure. You temporarily separate the
parties from one another, and you hear no more about the case. In
such examples, if the duration of the disease is prolonged, childbirth
will certainly cure, or almost certainly, because in childbirth you have
an imitation of a treatment (which is undoubtedly of value in these
cases) used in the case of the analogous disease attacking the anus.
Cutting through the mucous membrane, or deeper, and expanding
the anal orifice, cures the irritability and the fissure of the anus. And
so also in these parts. Vaginitis, a common cause of secondary vagi-

nismus, is generally easily cured. Chronic vaginitis is sometimes very difficult to cure.

The case I have to bring before you presents a good picture of one form of the disease, and of a course of treatment that has utterly failed hitherto. But I do not at all despair of this poor young woman being cured of this very painful and distressing malady. The case is as follows:

E. P., aged twenty-one, has been married for two years; is strong and healthy, and has menstruated regularly since she was fourteen years of age; has sexual appetite, but dyspareunia amounts now to complete impotence; has a slight yellowish discharge. She was admitted to Martha Ward, seeking relief from dyspareunia.

This condition has, in a case I have known, been made the ground of a divorce. You may conceive, therefore, what an amount of misery and evil may result to a woman from this disease when it amounts, as in this case, to complete impotence.

On examination there is found ulceration of the lower half of the end of the urethra, which is very vascular, and projects like a caruncle. Around the orifice of the vagina, and external to the hymen, are five rounded spots of apparently healthy superficial ulceration, of the size of one or two lines in diameter. They may be touched without producing loss of blood. The hymen is lacerated, and its posterior part is thickened, inflamed, and projects. The affected parts are intensely tender. No evidence of syphilitic or gonorrhœal affection is discoverable. The thickened portion of hymen was excised, and the five ulcerations were well cauterized by the thermo-cautery. Twenty-seven days afterwards it was found that three of the cauterized spots were healed; but anteriorly on the right side were two new little ulcers. There is now a small tubercle just within the margin of the fourchette. The lowest part of the posterior columna rugarum has become slightly hypertrophied. There is no ulceration of the urethra, which is now healthy. Dyspareunia as before. A month after this examination she, in my absence, came under the care of Dr. Godson, who dissected off the whole of the hymen, and made an incision through the fourchette—a proceeding which has been systematically recommended. After another month she declares herself as feeling better, but the dyspareunia remains as before. The urethra now presents only slight caruncular redness on the left side posteriorly. On the right side of the urethral orifice, and about half an inch distant from it, is a new speck of ulceration. On the right side of the vaginal orifice is another ulcer like the former, but somewhat larger; another posteriorly near the fourchette; and still another to its left.

Although treatment has been in this case successful, the success has been of a kind not to boast of, because it has always been followed by a reappearance of the disease. The woman is at present feeling better than when she came originally under our care, but she is still suffering from this curious disability.

Before concluding, I may tell you that in several cases of this ulcerative disease I have operated by excising the diseased bits, and without success; that is to say, the disease has reappeared after the parts affected were removed by knife. And what is extraordinary about these cases is this, that in other women you will have apparently the same disease, even much more, without any vaginismus, even without any pain. There are many women who have ulcerations (of which this case is a good example) who are quite unaware that they have any disease at all, who have no dyspareunia and no complaint. In a case of this kind which I saw lately operation by the actual cautery was, after a consultation, resorted to, and with complete cure of the disease, so far as the ulceration was concerned; but the woman has now around the orifice of the vagina several tubercles, which are red, not painful, and which indicate what I have already said is my own impression, that the disease is analogous to lupus rather than to eczema or any other disease with which I can place it side by side.

XII.

ON SPASMODIC DYSMENORRHŒA.

THERE are many kinds of dysmenorrhœa, some of which have little claim to the name. The most characteristic form of dysmenorrhœa is that which I have called spasmodic. A woman may be said to have dysmenorrhœa if she suffers from headache during the monthly period, or if she has sickness. In the same way she is said to have ovarian dysmenorrhœa if she has pain in one or other ovary during the monthly period. But that is not dysmenorrhœa proper. There are two chief kinds of dysmenorrhœa,—the inflammatory and the spasmodic. Spasmodic dysmenorrhœa is extensively known by the name of neuralgic; latterly it has been generally described as obstructive or mechanical dysmenorrhœa; these words "obstructive" and "mechanical" implying a theory of the disease which I shall speak of presently, and which I am sure is quite erroneous. This disease, called neuralgic, obstructive, mechanical, or spasmodic, is a disease of the nature of a neurosis, in which the contractions of the uterus cause great pain.

Contractions of the uterus are much better studied, for reasons that are plain, in the lower animals than in women ; the contractions particularly, of the unimpregnated uterus. From observation of them, and for other reasons, physiologists are agreed that there are contractions more or less regularly going on in the unimpregnated uterus of women, and especially in menstruation, whether healthy or morbid. The disease we are now considering is, in its essence, morbid contractions of the uterus occurring in connection with menstruation.

On the subject of these contractions I shall say a few words. In some conditions of disease, as in some uterine fibroids that are imbedded, the contractions are easily made out ; in other diseases, such as dysmenorrhœa, they are only believed to exist as the result of an argument. Some of the phenomena, which are explained generally by contraction of the unimpregnated uterus, are not due to contractions at all ; they are due to the pressure relations of the uterus—a

very difficult subject. For instance, if you place an intra-uterine
pessary or a tangle tent into the uterus, it is generally expelled if a
plug is not put into the vagina to keep it in its position; and this
expulsion of the tent or of the pessary is supposed to be produced by
contraction of the organ. It is very natural to suppose so, but I am sure
it is usually not the case. It arises from the condition of the woman's
uterus as to positive or negative abdominal pressure. You can easily
study this subject in any case in which you are placing a tent or a stick
of zinc-alum into the uterus or its cervix. You will find in most
uteri the tent or the zinc-alum slips out; but it is manifestly not on
account of contractions. Contractions are not brought on so quickly
and in a way so exactly in accordance with the repeated pushing in
of the tent. Besides, you will find many uteri in which the tent or
the pessary, instead of coming out, has a tendency to go in—an injurious tendency. Cases are not very rare in which a metallic pessary,
with a button upon the lower end of it to keep it in its place, is drawn
into the uterus altogether—button and all. I have seen this happen
several times; and considerable difficulty arises in removing it, when
it has thus got incarcerated in the uterus. These facts contribute to
showing that the phenomena we are speaking of are not caused by
uterine contractions; and I shall tell you another remarkable phenomenon which illustrates the same thing. The sticky cervical mucus,
as you are all aware, in 999 cases out of 1000 hangs out of the uterus
into the vagina; but I have seen it, instead of hanging out of the
uterus into the vagina, ascending, and filling the cavity of the body
of the uterus. This forms a good text, of great importance in pathology, which I hope to lecture upon some other day. This function
of the uterus when it acts in the way I have mentioned, drawing the
cervical mucus into the cavity of its body instead of expelling it,
certainly tends to produce morbid conditions of the uterus itself. The
same condition is illustrated in pregnancy. The ascent of the pregnant uterus itself is a phenomenon in this category; but during pregnancy, as I have seen in several dissections, the cervical mucus, instead of running into the vagina, ascends and runs into the uterus,
and hangs into the uterus instead of into the vagina; and this circumstance has led to considerable mistakes recently in the investigation
of the condition of the cervix uteri during pregnancy.

The best evidence we have of uterine contractions during menstruation is from the observation of cases of dysmenorrhœa spasmodica,
and this observation reveals that the contractions may be either clonic

or tonic. The clonic contractions are probably the most frequent. By "clonic" you know I mean come-and-go contractions like uterine pains. You will find women suffering from dysmenorrhœa tell you the pains come in pangs; and in the most violent pangs, in the most severe cases, the contractions not only affect the uterus, but may also affect the bladder and rectum, producing strangury and tenesmus, and also violent abdominal bearing down by reflected influence. Tonic contractions of the uterus, however, are not uncommon, and then you have the pain incessant, probably because the contraction is almost unceasing.

Some have sought for an analogy for this disease in urethral stricture. I shall mention two analogous diseases. The first is after-pains. You will often read in books that when a woman has after-pains there is a clot or a retained bit of placenta, or something which the uterus is attempting to expel; and this may be true, but such after-pains are not severe. That is not a disease—that is a healthy condition of the womb; the womb is doing its duty, as it were, and such after-pains are not very painful. The real disease of after-pains is a disease in which the recently emptied uterus goes into the most violent and painful contractions, without any discoverable object in view; and a severe case of this kind is a most painful disease, far more painful than the after-pains which come to expel a clot or a bit of retained placenta. Now, these violent after-pains are, I believe, connected not only with a morbid condition of the muscular tissue, but chiefly or primarily with a catarrhal condition of the mucous membrane covering the inside of the body of the uterus, a condition not without several analogies with the healthy menstruating uterus.

There is another disease not uterine, with which spasmodic dysmenorrhœa has an analogy,—spasmodic asthma. This is a disease affecting muscular fibres, and it is induced, as you know, in those who have a tendency to it, by the slightest catarrhal affection of the trachea and bronchi; and it is cured under a copious secretion from the mucous membrane; just as dysmenorrhœa is generally cured when the menses run freely. In healthy menstruation a woman has the mucous membrane of the cavity of the uterus in a catarrhal condition; it is not called catarrhal, because it is natural and healthy, while catarrh implies something morbid.

Spasmodic dysmenorrhœa may be combined with the exfoliative or membranous form; or rather, menstrual membrane may be discharged, the violence of the contractions separating and expelling bits that are

possibly otherwise quite healthy.   Such bits do not present evidence
of being separated by hæmorrhage into the middle of the mucous
layer, in adhering laminar clots.

Spasmodic dysmenorrhœa occurs at any age.   It occurs in women
otherwise most healthy.   It is specially liable to attack women at the
marriageable age; still more, women who, although married, are
sterile.   It is very liable also to attack women who have had large
families—we may call them excessive families; although, in such circumstances, the elderly woman makes less to-do about it, and does
not get for herself the same amount of sympathy as the young woman
does.   There is another set of circumstances in which it frequently
occurs, namely, when a fibroid is beginning to grow in the muscular
tissue.   If you find an elderly menstruating woman having persistent dysmenorrhœa, you should suspect that there is some growth
of this nature going on, and you will frequently find it verified in the
further history of the case.   Only a few minutes ago I saw a case of
this kind, where a woman, nearly forty years of age, began about two
years ago to have severe dysmenorrhœa.   She had seen several doctors of eminence, who told her that her disease was simple dysmenorrhœa, and I have no doubt they spoke truly as far as diagnosis could
go.   But now, after two years, there is a considerable fibroid in the
uterus, and there can be no doubt the dysmenorrhœa was started by
the growth of this tumor, which at first was too small to be discoverable.   Intense dysmenorrhœa, with fibroids of considerable size, is
also far from rare.   The disease I am considering is a disease that
frequently occurs in minor forms, especially in connection with unnatural or morbid conditions of the uterus, besides those that I have
mentioned.   For instance, recent authors say a great deal about its
connection with uterine displacement.   But dysmenorrhœa produced
by this cause is slight in degree, and is a symptom of the displacement, or of some morbid condition complicating the displacement.
This displacement has been a favorite cause with those who believe
that the dysmenorrhœa is mechanical or obstructive.   They say that
flexion of the passage obstructs the discharge of the blood.   Nothing
could be more erroneous.   There was recently exhibited to the Obstetrical Society the section of a uterus in the extremest degree of
acute flexion; and anybody who takes the trouble to look at that section will see that the flow of menses along that flexed uterus would
be obstructed only in a degree that practically cannot be of the
slightest moment—not nearly so much obstructed as the passage of

the blood along a flexed limb; not nearly so much obstructed as the passage of the water along a bend of the river Thames. Blood could run out through that model of an excessively flexed uterus just about as easily as if it were straight. In such cases the blood is said to be dammed up in the body of the uterus; and the uterus is described as thereby hypertrophied or dilated. I am satisfied that that is bad pathology. When you have dysmenorrhœa spasmodica accompanying real mechanical difficulty, then, as I have already said when speaking of after-pains produced by a clot in the uterus, or a retained bit of placenta, you have very moderate pain; you have not a fine specimen of the disease at all,—the dysmenorrhœa is trifling. This is exemplified in cases where you have truly mechanical difficulty, cases of dysmenorrhœa membranosa, where the membrane has to be expelled through the narrow channel. Well, in such cases, everybody knows the pain is slight compared with that of a characteristically severe case of the disease we are discussing.

Dysmenorrhœa spasmodica may occur at any time. The woman may have the violent pains of dysmenorrhœa apart entirely from ovulation or menstruation. In the majority of cases the pains begin before menstruation begins; in the majority of cases it is most severe just as the menses begin to flow; and, in the majority of cases, it diminishes as soon as the flow is free. It is seldom that a woman has violent dysmenorrhœa after the first two days of menstruation; for, within the first two days of menstruation, the quantity of the discharge has reached its highest. This fact, which is subversive of the mechanical theory, is familiar to women. Nothing is more common than for a woman suffering from dysmenorrhœa to tell you that she has most pain when she has least discharge,—that when, for any reason, the menses become scanty, the dysmenorrhœa becomes worse and worse; but, when the menses become abundant, the dysmenorrhœa is diminished.

Dysmenorrhœa not infrequently, even in the severest cases, disappears for one or two periods. In one of the severest cases I ever saw, a young woman in whom I was very reluctant to resort to mechanical treatment, the disease disappeared during her residence in Ireland for several months; it reappeared as soon as she came home to England. That fact, which I have seen illustrated in many other examples, is quite inconsistent with the popular theory of mechanical obstruction by stricture.

Still more about the theory of this disease. I have told you that it is a spasmodic disease, not an obstructive one, and if our knowledge of it is to be improved, it will be from studying, not cases complicated by flexion or tumor, or inflammation anywhere in the neighborhood, or in any part of the uterus itself, but by studying simple cases. And simple cases are abundant; they are no rarity. Simple cases are those where an examination discovers no additional morbid condition whatever. These constitute the majority. No disease, tangible or visible, can be discovered, and yet the woman has this violent disease near and during her monthly times. When examination is made with a view to find out that the case is a simple one, a uterine probe may be passed into the organ. As soon as it advances little more than an inch, it approaches the seat of the disease, the body of the uterus. In a healthy woman the internal os uteri and the whole interior of the body of the uterus are sensitive—that is to say, the touching of them by a probe is disagreeable. In a woman suffering from dysmenorrhœa spasmodica, the pain of touching the internal os is intense, and the pain is aggravated by passing the probe further on and touching the body and fundus; and in every characteristic case the woman at once tells you that that is the pain of her disease. The touching of these parts brings on the spasms, and the removal of the instrument may not be followed by arrestment of the spasms for a few minutes. It is in these simplest cases of dysmenorrhœa that the disease must be studied in order to discover its true nature and cure.

I have already said the disease is frequently complicated by uterine displacement and by uterine hypertrophy; but so far from these having anything to do with the most characteristic form of the disease, the worst cases occur in uteri that are ill-developed, uteri that are small. We have had an illustration of this in "Martha" lately—a case in which our treatment did little good to the woman's dysmenorrhœa. This woman had an ill-developed uterus, about two inches long, and acutely anteflexed. I must tell you of another case which occurred not long ago in Edinburgh, and which was seen by many physicians. In her, the dysmenorrhœa was of the intensest kind; but it was without any bloody loss at all. I at one time possessed this woman's uterus, and it measured only an inch and a half in length. Her sufferings were of the most intense kind; and, I may tell you, the most intense form of dysmenorrhœa constitutes one of the most severe and violent diseases that you will ever have an opportunity of seeing. The woman is, while it lasts, almost insensible, sometimes in a state of convulsion or spasm. She is cold, vomiting, and looks as if she were dying.

To-day I have not read to you any history of cases, because the reading of histories of these cases would not be, as in former lectures on other subjects, the filling up of a picture to give you a better idea of what we are speaking about. To go over all the details of cases of simple dysmenorrhœa would add very little to what I have told · you. The women may be in perfect health, except this.

Before I pass on, I must say a little more about the mechanical theory. In the various cases that have been in "Martha" within this last year we have found no stricture, no contraction of the passage through the womb, except in one case. Especially did we find no contraction when the woman was suffering from the pain; for in order to satisfy ourselves as to the nature of the disease in some of the cases, we passed a bougie into the womb while the woman was in the agonies of dysmenorrhœa, and we found that the passage was clear. This subject of a passage for blood I have not time to enter upon at length; it has been carefully discussed in scientific papers. I merely remark that the smallest passage described, "pin-point os uteri," as it is called, is quite enough to allow a hundred times as much blood to pass as there is any occasion for, or as offers to pass. Contraction, it is said, may be produced by swelling of the passage; but there is no special swelling of the passage, as may be found by examining in the way I have just described. Then another method of explaining the stricture is the blocking up by mucus or a blood-clot. But this kind of mechanical obstruction, even if it exists, does not induce severe dysmenorrhœa; it induces healthy uterine contractions, not of a very painful kind, fitted to force on the clot or the obstructing mucus. In an ordinary woman the cervix uteri gives passage to a No. 9 of the male bougie series. The bougies I show you here are just like the male bougies, only with a different curve. No. 9 generally passes a virgin's internal os uteri without any difficulty. This is important for you to know in connection with treatment. In the contracted cervix that I referred to, a No. 7 only could be passed at first. In treating a case of this kind you must find out what is the natural size of the cervix, in order to know how to adapt larger bougies to the case.

Now, how do you treat a case of dysmenorrhœa spasmodica? In the great majority you trust entirely to drugs and regimen; it is only in severe cases that you use mechanical treatment. Medicines for the treatment of this disease are not very efficient. Their great number and variety is a sufficient proof of itself that they are inefficient.

Those which are most valuable are laxatives (especially salines), diaphoretics (especially hip-baths and guaiacum). Lastly, there is the treatment by drowning the pain with narcotics and anæsthetics. A familiar treatment that mothers use, and often very efficiently, is well known. The young girl suffering in this way gets a hip-bath, a little strong gin-and-water hot, and is put to bed. She perspires and goes to sleep, and gets over the difficulty. But I cannot pass from narcotics without cautioning you, for social rather than for medical reasons, as to their use, especially the use of opiates. The disease is a chronic one; it is likely to recur every month for a considerable time, and you are in very great danger of teaching your patient the opium habit, which is a very much greater evil, and, indeed, a greater disease, than the other one you are curing. It is only in the rarest cases that you use opium, and recommend it to be used, systematically. In the immense majority of cases, even of those that may be called severe, if you are a wise practitioner you will say to yourself: "Rather the disease than teach my patient the baneful and almost incurable habit of opium-eating." In the course of my life I have known an immense extent of evil done by this prescription of opium for dysmenorrhœa—evil done not only to the patient herself, but to whole families: evil of very great degree.

Finally comes the mechanical treatment, and this treatment is very successful. I know no drug that can compare with this in its direct utility. I know very few treatments that are more decidedly useful than the treatment of dysmenorrhœa by mechanical means, and yet I recommend you in the great majority of cases of dysmenorrhœa not to resort to it. Dysmenorrhœa is a disease which occurs in virgins, and in them you will be most reluctant to use it. In married women who are sterile, you will be, on the other hand, easily induced to try the treatment, in the hope that you will not only cure the dysmenorrhœa, but also at the same time remove the sterility. In regard to the use of this treatment in virgins, I must say a few words in order to guide you as to when you are to resort to it. No rules that I can give you will make up for want of good sense and good feeling on your own part, but I shall give you some hints. The first is that you should, as a rule, not resort to this treatment in an unmarried young woman without the concurrence of three parties—firstly, your own approval; secondly, that of the mother or guardian of the patient; and, thirdly, that of the patient herself. All of these should be quite aware of the circumstances and of what it is proposed to do. Then

I believe you are justified in recommending it in cases—and they are not rare—where the woman's whole mode of subsistence is ruined. In one of the cases we had in "Martha" the patient insisted upon our doing anything whatever that was at all likely to relieve her, because she could not keep her situation as lady's maid, for she was confined to bed for three days every month by the disease. That was a sufficient reason in that case; and I can tell you that that girl was cured after a few days' treatment in "Martha," and came back to us to testify her gratitude for being able to keep her place, going about during her monthly period without letting her mistress know that she was ill at all. Then, in other cases, the general health is ruined; and this is not very uncommon. When a woman is laid up and prostrated by a severe attack of dysmenorrhœa every four weeks her health may gradually give way, and under such circumstances there can be no hesitation in resorting to the treatment. There is another set of cases where the severity of the pain is such as to leave no doubt as to the propriety of resorting to any means that offer a hope of cure; and cases of this kind, although rare, are still such as you will all meet with in the course of your practice. In some cases the severity is not so much in the pain as in accompanying phenomena. Lately, for instance, I had no hesitation in recommending mechanical treatment in a young unmarried female, not because the pain was extreme, but because when the pain came she had attacks of suicidal mania; and these attacks of suicidal mania were severe when the dysmenorrhœa was severe, and if the dysmenorrhœa was slight they did not come at all. Under such circumstances no one would hesitate to recommend the mechanical treatment.

Now, the mechanical treatment is very simple if carried on on the oldest of all mechanical plans recommended for the treatment of this disease,—that by bougies such as I show you here. The treatment by bougies I recommend to you because it is unaccompanied by danger. The only evil result I have ever seen from it is a temporary perimetritis. It is a treatment, the innocence of which arises from the fact that there is no cutting, and that the instrument is not left in the womb above a few minutes at a time. It is allowed to remain till the pangs of pain which it brings on have passed. In order to effect a cure you must go up considerably above a No. 9. You must go up so as to stretch and distend the internal os uteri; and this stretching or distension of the internal os may require you, in different cases, to reach different sizes. A No. 11 is quite sufficient in many cases; in others you will go up to a 12 or 13—rarely above that. These various numbers are not all used in one day, but in successive days, or every second or third day, and generally the whole is effected in a few sittings—say, from four to eight. You are not to expect that this treatment will cure every case. I can only tell you that most of the characteristic cases are, if not cured, at least greatly ameliorated. In several cases which have passed through "Martha" we have had failures, and we have had an ordinary amount of success. In one of them the success was remarkable: a single passage of the bougie through the internal os uteri seemed to be enough to dispel the woman's disease.

XIII.

ON HEPATIC DISEASE IN GYNÆCOLOGY AND OBSTETRICS.

A VERY striking case, recently in Dr. Southey's ward "Faith," attracts me to this subject of lecture. In gynæcology, and, indeed, in all departments of medicine, you will find a great deal of vague talk about the influence of the liver in producing or aggravating disease. This talk increases with the imperfection of the works in which it occurs. If you look to the best books of gynæcology and on diseases of the liver, you will see the least of this kind of remark ; and it belongs rather to medical lore than to medical science. The best authors are content to leave it more to tradition than to write it down solemnly in books. I am not disposed at all to deride this kind of medical lore, neither am I disposed to take up your time with it on the present occasion, because I have much more definite information to give you upon a very important subject.

In women the only specialty I have to call attention to, with regard to the anatomical condition of the liver, is that it is lower down ; in consequence of the peculiar shape of the chest, the liver lies lower in the right hypochondriac region, or at least produces dulness lower down, than in man. In examining the liver in women you have to take special care not to be misled by the results of tight-lacing ; the displacement of the liver—indeed, the deformity of the liver—sometimes produced by this is so great as to be very misleading were you not aware of its occurrence. Amenorrhœa has been described as being produced by fatty liver. I can neither confirm nor dispute this, which, so far as I know, is a mere assertion ; but I must add that I do not believe it. Fatty liver very frequently occurs in phthisical women, and in such you know that, for other reasons, amenorrhœa is common. That is a very different thing from saying the amenorrhœa is the consequence of this special lesion, fatty liver. Hyperæmia, or congestion of the liver, is said to be produced by suppression of the menses and by the menopause. I am not aware of anything that confirms this statement.   Portal obstruction, however, such as occurs in cirrhosis of the liver, might naturally be expected to produce congestion of the womb, as well as of the other pelvic organs, and menorrhagia ; and of this I have seen undoubted examples. One occurred not very long ago in "Martha." The woman was thirty-one years of age ; she had borne eight children, and had had three miscarriages, the last of which occurred a year before her admission into "Martha." Since that last miscarriage she became very ill with chronic hepatitis. Of the chronic hepatitis she was quite unaware, but simultaneously with the occurrence of the chronic hepatitis, her menses became more profuse and long-continued, and it was on account of this condition that she came to "Martha." We examined carefully the pelvic organs, and could find there no disease to account for the menorrhagia ; and being satisfied that her chief disease was chronic hepatitis, and that the chronic hepatitis was the cause of the menorrhagia, we had her transferred to Dr. Church's care. There I believe she soon died of the disease of the liver. Menorrhagia here was a distinct result of the hepatic affection.

I now come to consider the influence of hepatic disease in pregnancy. Here you will find remarks very common about the pressure of the gravid uterus upon the liver, disordering its functions and leading to disease of the organ. Even in good authors, statements like these occur. Similar statements probably you are familiar with in connection with the uræmia of pregnancy and parturition. I ask you, in the meantime, not to believe any such statements, neither with regard to the liver nor with regard to the kidney ; and I cannot omit here making a remark which is of very great importance in medical philosophy. You would scarcely believe, yet it is quite true, that men state, as if it were a fact, that there is great pressure upon the liver and upon the kidney in pregnancy, and proceed to reason upon this, not only to found theories of disease upon it, but also plans of treatment, and all the time they have never given even the slightest good reason for believing that there is any increase of pressure. Surely, the first thing in such circumstances is to demonstrate the increase of pressure ; but not one of the authors I have alluded to ever seems to dream that that is the first thing. You must first prove that there is pressure, before you proceed to found upon it as the cause of disease and as the basis of a line of treatment. There is no evidence, but rather to the contrary, that there is any increase of pressure upon the liver or kidney in pregnancy. Again, if you turn to the clinical view of the matter, and regard cases of large fibrous tumors and of large ovarian dropsies, you will see nothing to confirm the belief that pressure has anything to do with producing disease of the liver or of the kidneys; for in those cases, you might expect far greater pressure than in pregnancy, in consequence of the frequently far greater size of the tumors.   In pregnancy, diseased liver, however, is a matter of immense importance.   The liver has been, at least once, observed to be folded upwards upon itself in a case of tight-lace liver, the pushing up of the uterus producing this displacement, and fatal jaundice in consequence.   Rupture of the gallbladder has been observed in labor.   After delivery, hæmorrhagic softening of the liver has been observed, and a case has recently been put on record where, in connection with such softening, the organ burst or was ruptured, and fatal consequences ensued.

But now I come to the chief topic of the lecture, and I will begin with offering some observations with regard to one of the most important diseases of pregnancy, namely, persistent and uncontrollable vomiting.   Of the vomiting of pregnancy there are at least two kinds. There is what may be called the common kind, which, when severe, is almost certainly the result of morbid innervation.   Whether the sickness and vomiting of pregnancy is a reflected sensation, or a reflected motion, or the result of a reflected secretion, it is a consequence of morbid innervation.   It is frequently very grievous, and perhaps is sometimes even fatal.   This kind of vomiting in pregnancy is arrested when the fœtus dies.   It is arrested certainly by abortion, miscarriage, or delivery at full time.   It is not accompanied by any symptoms of grave disorder, except such as arise from deficient nutrition. But there is another kind of vomiting in pregnancy, our knowledge of which is extremely imperfect, and upon which some remarks are called for.   These are cases of vomiting in pregnancy, described by many authors, which prove fatal, sometimes suddenly and unexpectedly, without any apparent cause, or without any suspicion of the cause at the time the histories of such grave cases were written. Along with such cases have to be included some similar cases of sudden and unexpected death after delivery.

Our knowledge of the physiology of parenchymatous degeneration of the great glands seems to throw light upon this fatal or extremely dangerous form of vomiting in pregnancy, to show that the vomiting in such cases is something more than a morbid innervation, that it is

a symptom of a visible lesion. The case that I am about to read to
you I am quite sure, some years ago, I should have regarded as one
of vomiting in pregnancy proving fatal. I should not have known
how to go any further. But you will find, as I proceed, that I have
dealt with that case in an entirely different manner, deriving my
knowledge from recent researches into the nature of a disease called
icterus gravis, or what used to be called yellow atrophy of the liver.

It has recently been discovered by eminent physiologists that, in
health, the earliest stage of this disease is produced; that is, that the
great glands of the body (especially the liver) undergo in healthy
pregnancy and in healthy suckling a certain degree of parenchymatous degeneration. This is the first stage of the grave disease which
I have named. Like the watery blood of pregnant women, this
parenchymatous degeneration is not called a disease, because it is the
normal condition. If it were found in a woman not pregnant or
suckling it might then be called a disease; but, as it is believed to
be the regular normal condition, we do not call it a disease. No doubt
this parenchymatous degeneration is, so far as our shortsightedness
guides us, an extremely unfortunate thing for women, leading them,
as it were, upon the ice, and making them liable to dangerous diseases.
The condition of the blood is, probably, a chief part of the cause of
the proneness of women to disease of the kidney and uræmia; the
condition of the glands probably being the cause of their proneness
to further dangerous stages of parenchymatous degeneration, chiefly
of the liver, but also of the kidneys and other parts. Now, if you
look into the histories of fatal cases of vomiting in pregnancy, and
fatal cases of a similar kind occurring, just after pregnancy, in the
puerperal state, you will find slight jaundice often mentioned; you
will find, in many of them, hæmorrhages are mentioned as occurring; and a condition of lethargy, almost of coma, is described; and
these statements seem to me to make it almost certain that the conditions causing death were the result of the aggravation of this physiological condition of granular degeneration that I have been referring to. In order that you may further see how difficult it has been
to reach the truth in this matter, I must tell you that cases of this
disease occur without jaundice, or with very little, and without hæmorrhage and without convulsions; that is, without any of the ordinary
grave symptoms of the fully developed disease. I am presently to
describe a case. This concludes what I have to say about the dangerous and fatal cases of vomiting in pregnancy, and about the dan-

gerous and fatal cases of a similar kind occurring in the puerperal
state.

I can remember four cases of jaundice and slight albuminuria
coming on shortly after delivery. One of them also presented hæmatemesis. They had little or no accompanying fever. The occurrence
of such a combination of conditions has always been very alarming ;
but great prostration and the present danger of death have occurred
in only two of the cases. That which had hæmatemesis presented no
constitutional disturbance, except slowness of pulse and a temperature
less than normal.

But before I come to the special case of to-day's lecture, I must
say a few words upon jaundice occurring in pregnancy. Ordinary
jaundice rarely occurs in pregnancy—jaundice from obstruction, or
from catarrh of the stomach and duodenum. I have seen pregnancy in
a woman who had a chronic jaundice, and I have seen jaundice come
on during pregnancy. In regard to this kind of jaundice there is
very little to be said. You cannot mistake this disease. The name
of the disease implies all that is necessary for its diagnosis. Anybody
can tell when a woman is jaundiced. It is not a mere tinting, but it
is, as the disease you are all familiar with, quite easily recognized.
This disease occurring in a pregnant woman does not make her very
ill—at least, not more than it would if she were not pregnant ; but
the woman having it runs considerable risk of abortion or miscarriage.
And, when this occurs, the abortion may be directly the result of
the jaundice, the child being born alive, and, if the disease has not
lasted long, untinted by the jaundice ; or the jaundice may kill the
fœtus, and the abortion or miscarriage in that case may be a secondary
result of the jaundice—the result of the death of the fœtus, not of
the jaundice directly—and then the fœtus and all the membranes are
deeply tinted with the coloring-matter of the bile. Now, in a case of
this kind you may have to consider the importance of bringing on
premature labor ; but this should only be done if the disease is intense and long-continued, and if the child is alive. No doubt it is
also worthy of your consideration whether you should not induce it
in some severe cases, from the fear of the supervention of the icterus
gravis as a consequence of the ordinary jaundice. It is impossible to
lay down rules with regard to this last point, because cases have not
yet sufficiently accumulated to form a basis of experience for such
rules. I must therefore leave it in this undecided form.

Icterus gravis, or the yellow atrophy of the liver, is a rare disease,

and has only been well recognized within a generation or so; and I
have no doubt that our increasing knowledge, of which I have tried
to give you a sketch, will lead to its being found to be not so rare a
disease as has hitherto been supposed. Especially will this arise from
what is now known, that the essence of the disease may be there without the presence of all or even of any of its grand indications during
life; and its grand indications are convulsions, jaundice, and hæmorrhage. If uræmia from disease of the kidney occurs once in about
every 500 women that are in advanced pregnancy or parturient, this
disease has not been observed oftener than once in 10,000. The disease is called yellow atrophy of the liver. There may be no atrophy
of the liver, the disease proving fatal in an early stage, as in the
case I shall read to you presently. The disease has been called
cholæmic eclampsia, just as the corresponding disease of the kidneys
is sometimes called uræmic eclampsia, from the frequency of the
convulsions. But there may be no convulsions in either disease;
and in the case I am to read to you there were no convulsions.
Hæmorrhage from the stomach or bowels or womb, or into the
tissues, is a very characteristic phenomenon of the disease, and yet
there may be none of it. In the case I am to read there were no
hæmorrhages. Lastly, the disease may be without jaundice; and
generally, as in the case before us, the jaundice is slight. Here the
jaundice got less as the woman got worse, instead of getting greater.
The jaundice is not like that which you know familiarly as the common jaundice; it is a much slighter condition of tinting, and, in the
cases of icterus gravis I have seen, never has proceeded to be a deep
yellow. The disease should not be called jaundice or icterus at all;
it is a disease which affects the whole body, and whose best-known
manifestations are in the liver. There you have not only the parenchymatous degeneration of the hepatic cells, which I told you was a
physiological condition in pregnancy and suckling, but further steps
of degeneration, which this is not the place to describe, going on to
complete fatty destruction of the hepatic cells. This, indeed, should
be called, if we only knew what the poison was, a case of poisoning,
perhaps blood-poisoning. One German author ascribes the disease
to poison from decomposition of the fœtus; but for this view he advances no argument except the analogy of other poisons. Believing
it to be a poison, he merely fixes upon this one, apparently without
any reason. Now, as I go on I shall, I think, satisfy you that it is
probable that instead of the dead and macerating fœtus poisoning the

mother, it is the mother's condition that poisons the child. A great author has also suggested that the disease is essentially uræmic; and, no doubt, the urea in the urine is very much diminished in this disease; but the disease is not at all like the ordinary uræmic eclampsia. Yet, it is true, parenchymatous degeneration of the kidney is found along and corresponding with parenchymatous degeneration of the liver.

Here I would mention to you an interesting set of facts in connection with this subject. If you read over cases of heart disease, especially mitral regurgitation, you will find that in them women are very likely to miscarry, and miscarriage is in them almost certainly a direct result of the disease. If there is a poison in the woman's blood, in this case it is probably a poison from imperfect aeration of the blood, and that induces the miscarriage. This has been almost proved by experiments on the lower animals, showing that the blood of dyspnœa induces emptying of the uterus. You have further evidence in the fact that the children are almost always born fresh, if not alive. The disease has brought on miscarriage; it has not killed the child. The condition of the blood has stimulated the uterus to action directly. In the comparatively common disease of the kidneys observed in pregnant women with albuminuria, you have an intermediate set of results between those of heart disease and those of icterus gravis. In uræmic patients miscarriage is not very common. There does not seem to be a great tendency to it. There is a tendency to it, but not great; and so far as I can form an impression from extensive experience and reading, the child may be alive. It is also frequently dead, and we know that in this disease the child may be killed by the uræmia. The uræmia that is part of the cause of the woman's disease is also found in the child. When you come to icterus gravis, however, you will find a different set of results. Not only is the child almost always born dead, but it is almost always born decomposed, and there seems to be no tendency to abortion or miscarriage directly. The uterus is not prone to throw off its contents; if it does throw off its contents it is as a secondary consequence of the death of the child; and the death of the child here has been shown to be the result, or, at least, to be connected with the presence of poisoning of its blood by the biliary acids which have been discovered in the fœtal blood as the cause of its death. There seems to be in the icterus gravis rather a tendency to avoid miscarriage. The fœtus is in most cases described as being macerated, sometimes putrid; and instead of abortion being induced, we have missed abortion.   The case I am to narrate to you is a case in which the womb, as it were, refused to throw off its contents, instead of, as in heart disease, being stimulated to throw off its contents.   Here there seems to have been the opposite tendency, and the dead and decomposing fœtus remained in it long after it would do so in ordinary circumstances.

Now I dare say you will be prepared for my telling you that there is little to be said about the treatment of this very important disease. Emetics, purgatives, and diuretics have been tried, besides other medicines.   The only thing I can suggest in the way of treatment is that the uterus should be emptied—this is with a view of saving the mother.   You may say, "Is there any chance of saving the mother in a disease like this?"   The impression abroad in the profession is that this is a necessarily fatal disease; but there are two reasons for hope: the first is, that we know that the physiological condition, the early stage of this disease, does no harm to a woman; the second is, that there is considerable probability that cases are cured by the death and expulsion of the fœtus, whether it happens spontaneously or is brought about artificially; and this appears to have occurred in the case I have to read to you.   So far as the history can indicate, we have reason to believe that this woman suffered from this same disease in her first pregnancy that proved fatal in her second; and probably some of the cases of dangerous vomiting that have been cured by abortion have been in the same category.   The present case did not occur under my care, but under that of Dr. Southey, who called me to see it in "Faith," and to him I must express my gratitude for the opportunity of observing and assisting so interesting a patient.

E. C., aged thirty-four, said to be of temperate habits, was married about a year ago.   Three months after marriage she had a miscarriage—after, it was supposed, the second month of pregnancy. At this time her condition was described as resembling that at the time of her admission to "Faith," only the jaundice was believed to be greater.   Her present illness began about five weeks before admission (December 2d), with vomiting and headache, of which the former has continued ever since.   She has kept her bed for three or four weeks.   The jaundice is said to be deepened in color.   She has had wandering delirium, especially at night.   On admission is in a wandering, dreamy state, and says she has no pain; is generally, but only slightly, jaundiced.   Tongue moist, not furred; breath offensive. No itching nor yellow vision.   Pulse 108; respirations 12; tempera-

ture 98.6°.   Fulness and supra-pubic dulness in the hypogastrium,
where there is also slight tenderness.   Does not permit a sufficient
vaginal examination.   Hepatic dulness normal; splenic dulness normal.   Urine dark-colored, bile-tinted; specific gravity 1012; turbid,
acid, albuminous (one-fifth); contains casts, epithelial and blood-
cells.   Takes milk and beef tea, but vomits almost everything.   December 4th. To have a borax wash for the mouth ; the bowels to be
opened by saline draught.   5th. Hiccough occasionally.   6th. A dark,
lumpy motion of bowels.   7th. Headache.   8th. Has slept well, after
taking ten grains of chloral.   9th. To be fed per rectum.   10th. Jaundice diminished; says she feels better.   Pulse 86 ; temperature 97.4°.
11th. Urine one pint and a half in twenty-four hours, albumen
(one-eighth); hiccough.   18th. Pulse 128; respirations 18; temperature 97°.   Purple discoloration of inner sides of thighs.   19th.
Tongue furred.   Is more drowsy and wandering ; vomits her food
mixed with bile.   Only a trace of albumen in the urine, which contains crystals of leucin.   Urea about sixteen grams in twenty-four
hours.   21st. The liver dulness slightly diminished ; jaundice less ;
quite rational when refusing to permit a vaginal examination ; ob-
jects to induction of abortion.   23d. Tangle tent introduced into cervix uteri ; urine runs away in bed ; tent removed after sixteen hours.
24th. Probe passed into the uterus, and a large tangle tent placed in
the cervix, with a sponge in the vagina ; ergotin to be injected sub-
cutaneously.   Died in the afternoon.   Post-mortem forty-three hours
after death.   The surface of the uterus, which is of about the size of
a cricketball, is congested, and so are the neighboring coils of intestine.   The liver small, weighing 2 lbs. 2 ozs., very soft and flabby
to the touch; its surface partly green (especially round the edges),
partly brown, with the lobules very distinctly marked ; no evidence
of congestion.   Gall-bladder contains healthy-looking bile.   Liver
on section yields an emphysematous feeling; color uniform, greenish-
brown at first, but becoming darker ; no trace of lobules to be seen ;
highly emphysematous (not putrid), its section resembling that of
highly aerated bread.   Spleen very dark in color and emphysematous.   Kidneys flabby, with large air-blebs under the capsule and
air-vesicles on section ; structure very indistinct, but presents evidence
of congestion of cortex and bases of pyramids.   The cavity of the
uterus contains air and shreds of membrane, and a much-decomposed
fœtus of about six weeks.   (She was held as being three months preg-

nant.) Placenta adherent, about two inches in diameter. Contents of bowels stained with bile. Stomach congested and presenting internally numerous air-vesicles. Left common iliac vein contains air with fluid blood.

These are only some of the details of this very important case. The microscopical details I have omitted altogether. It only remains for me now to offer a few remarks about the diagnosis of this disease from uræmia. This disease is comparatively a chronic one, occurring in pregnancy. Uræmia, or the disease of the kidneys connected with the uræmia, is generally an acute disease, running a rapid course and occurring most frequently during parturition. In this disease you have delirium, muttering, and lethargy rather than coma—conditions which are very different from the silence and deep comatose condition of a woman suffering from uræmic eclampsia between the fits. In this woman, and in cases of this kind, jactitation and restlessness are described. The reverse is the case in the coma of uræmia, and in our patient at the worst there was a possibility of rousing to clear intelligence for a few minutes, which is not observed in uræmia. In this disease you may, as in her case, have almost constant vomiting. Whether it is accompanied with sickness or not I am unable to say, but there was in this woman, even when she was not vomiting, the constant or very frequent going on of the efforts of vomiting. This is not observed in uræmia: violent vomiting for a time is not uncommon in uræmic cases, but it is only for a time; constant vomiting is not observed. In this disease there was observed great duskiness of the skin, and a peculiarly injected condition of the venous capillaries in the thighs. Now, in the deep coma of uræmic eclampsia you have simple cyanosis of varying degree, sometimes very intense; but you have not the duskiness and local blueness I have mentioned as occurring in this disease. There are other distinctions between the two diseases founded upon examination of the discharges from the body, especially upon the examination of the urine, and the two diseases are quite easily distinguished post-mortem. I have made these remarks upon the distinction of the two diseases because some authors have regarded the eclampsia and coma in both as the same; and, indeed, as I have already told you, at least one great author describes the coma and eclampsia as owning the same cause. The clinical history of the disease is very distinct, and shows no close alliance between them at all. I must warn you against supposing that anything I have said in regard to icterus gravis is conclusive.   The disease is rare, and has not yet come to that degree of distinctness of recognition that enables a lecturer to speak with precision and dogmatically ; but I feel quite sure that the subject is of such intense importance as to be well worthy of the time I have given to it.

XIV.

FIBROUS TUMOR OF THE UTERUS.

THE subject of this lecture is two cases of uterine fibroid which have recently passed through " Martha " Ward.

This disease, uterine fibroid, is one which has itself suffered from a very serious affliction, the disease of many names, very important from a student's point of view. The regular name is fibrous tumor of the uterus. The term uterine fibroid has been lately coined ; it is shorter, and on this account may supplant the old name.

Uterine fibroid is a disease of the childbearing period of life, not of any other ; a disease affecting the elderly woman rather than the younger during this period, and probably attacking women who are fertile rather than those who are sterile. It is a disease which affects the middle layer of the wall of the uterus alone. The chief constituents of the middle layer of the uterine wall are unstriped muscular fibre and connective tissue ; and these tumors generally consist of both of these structures in various conditions of development ; sometimes, however, of one or of the other almost exclusively. Its vascular constituent structures may be little developed or immensely developed. A fibroid may be telangiectatic—that is, the venous sinuses of the tumor may have a peculiarly great development. It may also be lymphangiectatic, which signifies a great and peculiar development of the lymph channels.

This tumor may develop itself wherever there is tissue of the kind constituting the middle layer of the uterine wall ; for instance, in the round ligament, broad ligament, or the Fallopian tube, or vagina.

The subject of to-day's lecture is ordinary characteristic uterine fibroid. This may grow in the midst of the tissue composing the middle layer of the wall of the uterus, in which case it is called imbedded, or intra-mural ; or it may grow on the outside of the middle layer, when it is called subperitoneal ; on the other hand, it may grow on the inside of the middle layer, in which case it is called submucous. This is a very elementary and incomplete statement of the three positions.

The cases upon which I lecture to-day are of the commonest variety—imbedded, or intra-mural. These are almost always more or less distinctly separable from the tissue of the uterus, surrounded by a capsule of less dense tissue than its own or that of the uterine wall, and in this capsule are developed enormous uterine sinuses. It is this great development of uterine sinuses around the tumor which gives it its chief importance. This development is analogous to that which takes place in pregnancy.

Now, what is the importance of these tumors? Why is it that they are of such intense interest to practitioners? Because they are very common. Because they are sometimes very large. Because they sometimes give rise to diagnostic confusion and difficulty, especially when they are complicated. When complicated with a cyst or chamber in their substance full of fluid, they are very liable to give rise to error in diagnosis. Such a tumor is called fibro-cystic, and is often difficult to distinguish from ovarian cystoma. When fibrous tumors of the uterus are complicated with pregnancy, the one or the other condition alone may be recognized, in which case an error of omission occurs.

I now come to the great interest of this disease. It is for the most part a bleeding disease, and might be called by the name of metrorrhagia. This would be giving it a nosological title, such as many diseases still have or retain. It is better, however, to adhere to the more distinctive and more scientific name,—uterine fibroid. It is true that there is sometimes no hæmorrhage, even amenorrhœa; but this is exceptional. The bleeding is frequently of a passive nature, or a more or less copious oozing, resembling that of a menstruation, and the loss of blood may be large, because the area from which the blood flows is often very great compared with the bleeding area in a healthy menstruation. Frequently, however, it is not a passive discharge, but a regular flooding; and in this case a woman bleeds as in phlebotomy, a large vein being laid open, and I have seen such openings. This kind of bleeding, I believe, leads to death, directly and indirectly, nearly as frequently as post-partum hæmorrhage causes death directly. Sometimes it causes death directly or at once, but more frequently indirectly, by producing extreme anæmia; the woman dying, perhaps, without any loss of blood at the time.   Examples of both of these fatal terminations are not very rare.

These tumors are further important, frequently interfering with utero-gestation.   They are themselves liable to disease; inflammation and sloughing, and probably other forms of degeneration.   They are dangerous to life from sometimes producing peritonitis; occasionally appearing to produce what is called cancerous peritonitis; the latter half of the name being suggested by the rapidity of its progress. Very rarely peritonitis and death are induced by peritoneal rupture. I have known peritonitis and death caused by crackings of the outside shell of a fibroid which was calcified *en coque*, and in which the *coque* was made to crack or burst outwards by the shrinking of the internal parts of the tumor.

Sometimes the disease produces an extreme and even dangerous amount of constitutional or gastric irritation, so called from our present ignorance of its real pathology.   Sometimes, but rarely, it causes obstruction of the bowels.   We have in our second case for to-day's lecture an example of a rare fatal termination of this disease, in the midst of convulsive and other nervous phenomena induced by uræmia, the consequence of partial and long-continued obstruction of the ureters.

Such, gentlemen, is a rough outline sketch of the pathology and formidable character of the disease of which we have recently had two examples in "Martha" Ward.   Now for the history of the first case:

J. G., aged forty-five, married twenty-three years, three children, the last seventeen years ago.   Catamenia commenced at fourteen years of age; last occurred a fortnight since; latterly irregular; interval from four to six weeks, of about four days' duration; on the last occasion the loss excessive.   No leucorrhœa.

Two years since, first noticed a swelling in the lower abdomen, which has been getting gradually larger; has no pain when still, but feels discomfort in the lower back when walking about; is extremely anæmic, with puffy swelling of the face.

Belly prominent, with uniform surface, semi-globose; lower half occupied by a firm elastic mass, with an indistinct feeling of fluid.   This mass moves freely from side to side; is not tender; no fluctuation; mass dull on percussion; resonance commences one inch above the umbilicus; per vaginam, cervix uteri much elevated; brim of pelvis presents fulness, but is otherwise natural; cervix healthy, with a minute excrescence seen to project from its interior, not felt by the finger. Probe passes easily to the left, and runs up on the left side of the tumor; its point can be felt a little above the level of the umbilicus, and about five inches to its left, when it has passed in six and a half inches; uterine souffle plainly audible in the region of the right iliac fossa.

Here was a very simple case, but I shall show you it might have been a very difficult one to a beginner. There were three small polypi in the cervix uteri from which the hæmorrhage might have proceeded, but probably did not do so to any important amount. They were removed by forceps. This tumor might readily have been mistaken for a six months' pregnancy; there was a round swelling, firm and elastic, and on manipulation it could be felt distinctly to contract. On auscultation, the uterine bruit or souffle was heard. Mark how erroneous it is to call this sound the placental bruit, and yet it is a term frequently applied to it.

The tumor was movable; the cervix was high, large, and soft. There were, therefore, in it many of the chief features of pregnancy, for the bleeding might have been referred to the mucous polypi. The age of the woman, and other points, however, were sufficient to dismiss such an idea in our immediate case; and so the uterine probe was introduced, which revealed a large uterine cavity (as would also be met with in pregnancy). The treatment of this case was not to produce absorption, such an occurrence is so rare that it must not be expected, it may be hoped for. Of course we removed the three small polypi, which had little to do with the hæmorrhage, and hæmorrhage was the only important symptom or condition to combat.

The treatment adopted was by ergot. In many cases this drug is of no avail, but in this case it had results so immediate as to strike me with astonishment. I must remind you that the tumor was so soft as to give the idea of fluid; this is exactly the kind which is known to be most benefited by ergot. We injected three grains of ergotin underneath the skin; this was repeated several times at intervals of a day; but it had at length to be discontinued, because it produced very serious diffuse inflammation of the cellular tissue, narrowly and fortunately without a suppurative termination. This effect we found to be peculiar to the patient, for the same injection caused no inflammation in several other women, the same solution and syringe being employed. In place of it, a fluid drachm of the liquid extract of ergot was given daily. The result of the treatment has, as far as I know, never been surpassed as regards rapidity of diminution of the tumor. The dulness which extended one inch above the umbilicus was in forty-eight hours reduced so as to extend only to the level of three inches beneath it. Such a remarkable result could only have been produced in a soft tumor.

This improvement was accompanied by arrest of bleeding. After being in the hospital two months she went out, having lost the puffy anæmic appearance, and having acquired a healthy aspect. The use of the ergot may now be given up, for a time at least. [Seven months afterwards she was heard of, and fully maintained all the improvement.]

A few words about abdominal tumors connected with the uterus, which diminish. You must not suppose that uterine fibroids are the only ones. From them, as the result of their shrinking, blood and œdematous fluid, which is often very abundant, may be expressed by the contractions of the muscular envelope. Indeed, at length, and probably also from the mechanical compression, the very tissue of the tumor may be absorbed. I saw a case in the hospital the other day in which there was a tumor in the lower abdomen, with loss of blood, while the patient was taking styptic medicine. I diagnosed a morbid pregnancy. The tumor rapidly diminished, and therefore the diagnosis was thought to have been very far wrong; but suddenly a dried-up placenta and fœtus were expelled. The liquor amnii had become absorbed, and this was the cause of the shrinking in this case. A hæmatocele may also rapidly disappear. You will remember a case lately in "Martha."

I come now to the second case, and with it I shall be brief.

A. S., aged forty, married eleven years; never pregnant. Catamenia commenced at twenty-one years of age; last menses appeared nine weeks ago, continued for thirteen days very profusely, accompanied by severe pain in the lower back and abdomen. The periods have generally been irregular; formerly the interval extended from two to six months. Two years ago they became regular; a scanty loss every three weeks. Six months later the periods became profuse, with only a fortnight's interval. Latterly there has been again a longer interval—four or five weeks—and the loss has been inconsiderable.

Complains now of a stabbing pain in the lower part of the belly, especially in right flank, shooting down the right thigh, coming on every few hours. Loss of appetite, constipated bowels, painful micturition, alleged great flow of urine.   Urine examined; almost color-less; reaction acid; no albumen.   Sp. gr. 1002.

The abdomen is prominent, semi-globose in shape, occupied by a dense hardness, said to be of twelve months' duration; the most prominent point is midway between umbilicus and symphysis pubis, the belly here measuring thirty-two inches.   The whole of the promi-nent part is very hard and elastic.   Hardness, dull on percussion, up to one inch above the navel.   No impairment of resonance elsewhere.

True pelvis nearly altogether occupied by a hardness, which can be identified with the abdominal tumor above described.   The vagina is natural, and contains a white discharge.   Probe passes easily into the uterus to the left, to the extent of six and a half inches.   Tenderness on both sides of the uterus, especially on the right.

The chief interest in this case lies in the remarkable way in which the disease produced a fatal result.   While in the hospital, undergoing treatment, she was seized with uncontrollable vomiting, which lasted for about eight days.   At the end of this time she began to have fre-quent and incessant twitchings, and at least twice actual convulsive fits.   She had also what I never saw before, a very remarkable lim-ited herpetic eruption upon the perinæum and posterior parts of labia majora; nowhere else.   This came on suddenly, and disappeared almost as quickly.   I have no doubt this was due to the morbid nerve influence which caused the twitchings and convulsions.   We were puzzled beyond measure at this unusual and unexpected group of phenomena.   The post-mortem, however, explained it all.

Post-mortem, thirty hours after death:

Abdomen.—Stomach distended; was not opened, neither were the intestines.   No peritonitis.   Evidence of old peritonitis, upper sur-face of the liver and spleen being adherent to diaphragm.   Liver: upper surface adherent to diaphragm, and the lower surface to upper border of right kidney; on section, healthy.   Gall-bladder full of light-green bile.   Spleen: adherent to diaphragm; healthy.   Right kidney: small, wasted; capsule comes off with ease, leaving surface smooth, pale, mottled with a few bloodvessels; cortex narrow, white; pyramids pink; ureter much dilated and tortuous, nearly as big as to admit a finger.   Left kidney and ureter same as the right.   The two weigh 11 ozs.   No clots in inf. vena cava, in right spermatic vein, or left spermatic vein.   At the junction of the internal iliac with the external of the left side is a large clot of a fibrinous nature, not completely organized, filling up the cavity of the vein.   On following the branches of the internal iliac vein a vein was found coming from the spine, and emptying itself into the internal iliac, completely blocked by a nearly organized clot. Filling up the whole of the upper pelvis is a large fibrous tumor, weighing 4 lbs. 10 ozs., and pushing the bladder over to the left. There is a large vein on the right side of the tumor, dilated, tortuous, and empty. Tumor hard and pale, no bloodvessels being seen in its substance. Cervix obliterated. The cavity of uterus is on the left, and is very elongated and dilated, but very pale. Right ovary pale, and displays a recent ruptured Graafian vesicle. Bladder: mucous surface very pale.

Now, how did this tumor produce the fatal result? It was very hard, and was jammed into the pelvis, and compressed the ureters. These ducts became greatly dilated and tortuous. The kidneys were irritated, and their structure became diseased. The nervous phenomena which preceded death were almost certainly uræmic; death being produced by the compression of the ureters as the first link in the chain of fatal consequences of the tumor. The urine when examined presented, as its only morbid condition, a low specific gravity, and this did not excite suspicion of the disorder that existed. Several times, when she was very ill, we wished to examine it; but, as it was passed in bed, none could be collected for this purpose. As I have said, we never suspected this lesion, and consequently were unlikely to diagnose it. In similar circumstances in future, besides looking narrowly to the urine, I would attach importance to pain in the flanks and down the thighs.

The urgency of this case was quite as much in the pain as in the bleeding; and it appeared to me that the pain might be diminished by relieving the great tension in the neck of the womb. The os was slightly opened, and the cervix very much on the stretch; the tumor growing down into its lip on one side. I incised it, therefore, with a pair of scissors, partly to relieve tension, and hoping to reduce the hæmorrhage, which the opening up of the neck in such cases seems sometimes to do. We contemplated also possible removal of the tumor by enucleation, but the autopsy showed that such a result could scarcely have been produced. The operation of enucleation would have probably proved a failure, and would probably never have been attempted, from the failure of the indications for its fulfilment, which should have been manifested in the progress of the case.

But the destruction of the tumor might have been attempted by other methods, such as by means of applications of the actual cautery.

And while the autopsy showed that successful enucleation could scarcely have been effected, it also showed a lesion of the urinary system which rendered removal of the tumor necessary for the saving of the woman's life from the kind of death which carried her off.

Time will not allow me to say more, but I shall have occasion to take up the subject of fibrous tumors of the uterus again, as cases of different kinds come under our notice in the wards.

XV.

CANCER OF THE BODY OF THE UTERUS.

THE subject of this lecture is cancer of the body of the uterus, a disease forming part of a great class of diseases,—cancers of the female genital organs and their neighborhood,—in regard to which a great deal has yet to be made out. The pre-eminently glandular organ, called the neck of the womb, is the most frequent seat of cancer in the female genital organs, but this pre-eminence is very much exaggerated. This arises partly from the fact that, as cancers in these parts go on, the neck of the womb becomes involved, and then the case—diagnosed as most cases of cancer are, in a late stage—is put down as a case of cancer of the neck of the womb, whereas really nothing is known as to where it originated. Lately, in "Martha," we have had thirty-nine cases of cancer in the interior pelvic region, and of these nineteen, or about one-half, have been put down as cases of cancer of the neck of the womb. But, even with regard to these nineteen, we have not invariably been certain that the disease ought to be so classified. We were sure that in each of these cases there was cancer of the neck of the womb, but whether the disease commenced there (and it is from the position of its commencement we would name such a disease) we could not tell. Besides nineteen cases of cancer of the cervix, we have had five cases which have been entered as cancer of the vagina; we have had four cases entered as cancer of the body of the uterus; we have had one case of cancer of the rectum; and we have had ten cases which have been classed either as cases of pelvic cancer or as cases whose origin was not only unascertained, but unguessable. In a former lecture in this course I described to you a case of cancer commencing in the sacrum, osteo-sarcoma. Cancer may commence in any part, and before I come to the proper subject of the lecture I shall say a few words about an interesting case, an example of disease which probably began in the rectum, but, as you will see, now affects the uterus as well.

E. W., aged thirty-five, was admitted March 10th. She has been
twelve years married, and has had four children, the last three years
ago, and she has not been in good health since that birth. The catamenia have been regular till six months ago; since then she has almost constantly lost some blood, and there has been at times a yellow
discharge. Complains of pain in the lower part of back, and in both
iliac regions, especially the left. Passes urine generally with fæces.
The latter are passed twenty times, or oftener, daily, and with severe
tenesmic pain, and with griping in left iliac region. The disturbance by her bowels is very annoying during the night. The sister of
"Martha" estimates the quantity of moulded fæces that is passed in
a day as a full ordinary amount, or rather more. Examination of
the abdomen finds nothing abnormal except a distinct doughy feeling
in the flanks and lower belly, evidently produced by accumulated
retained fæces. The whole upper part of the pelvic excavation, as
digitally examined per vaginam, is a hard mass, with deep fissures
diverging from what is taken to be the situation of the cervix uteri,
which cannot itself be identified precisely. This hard mass is only
slightly displaceable upwards and downwards. The discharge is thin,
blood-stained, and not fetid. The rectum, as felt per vaginam, presents a hard rounded mass, as if it contained a scybalum of the size
of a hen's egg. The finger, passed per anum, after permeating a
pouch about one inch and a half in diameter, reaches a tight stricture
in the seat of the egglike swelling. It admits only the tip of the
finger, and is situated in the midst of extensive fixed hardness.

This case presents an example to you of an accident which is not
common in the diseases of women, except in cases of cancer. It is a
curious fact that an ovarian tumor, a fibrous tumor, a pregnancy,
seldom cause great retention of fæces. When you examine some
cases, as, for instance, two women with fibroids at present in "Martha,"
you would think it was impossible for fæces to get past the hard tumor jammed into the brim of the pelvis; and yet it is the fact that
rarely do you see obstruction of the progress of fæces—such as you
see here. Besides malignant disease, as in this case, the scybalum
causing obstruction of the rectum is the most important cause of great
retention of fæces in women. This is not extremely rare; I have
seen it the cause of very great mistakes. In that case a woman passes
liquid fæces round the scybalum; and the case may go on even for
years, never passing a proper motion, the fæces always escaping in a
semiliquid form. That is not the case here. Here the fæces are

positively retained, and are not scybalous; there is no feeling of round scybalous masses, but you feel the woman's belly is really stuffed with semisolid fæces.   In this case you will have noticed that we look forward to performing an operation for the relief of the patient's sufferings.   Her sufferings are intense from tenesmus, accompanied by actual griping pain of a different kind from the disagreeable feeling of tenesmus.   This relief we expect to be able to give her by colotomy.   We propose colotomy in this woman because she is suffering a great deal, and because she has, so far as we can judge, the prospect of a considerable span of life yet,—I mean a span of life not measured by years, but by a considerable number of months,— and it is surely worth while to let her have the imperfect relief which is afforded by colotomy.   But on this I am not going to say anything more to-day.

Before I pass from this subject I wish to point out another very important practical fact, that while retention of fæces is frequently due to malignant disease, retention of urine (and of this we have illustrations at present in "Martha") is a disease rarely accompanying malignant disease.   Retention of urine is common in cases of fibrous tumor of the uterus; it is not common in cases of swellings, however large, produced by malignant disease.   I may mention that lately we have seen urinary retention in a case of cancer affecting the vaginal orifice, and mechanically impeding the exit of urine.

You will notice that when I enumerated cancers of uncertain origin in the pelvis as ten, we called a good many of these pelvic cancer; and I wish to point out what is extremely well illustrated in one case in "Martha" at this moment.   In that case the whole brim, the whole upper part of the excavation is a solid mass; and when cancer of the neck of the womb is not present, you have, if the woman is young, a very difficult diagnosis.   Now, what disease is there which is not at all uncommon, which is sometimes chronic, and which makes the whole roof of the pelvis, as in the old woman now in "Martha," like a board?   It is chronic perimetritis.   Some cases are quite easily diagnosed, but some are extremely difficult to diagnose; and I have often told you that, when you hear of a diagnosis being difficult, difficult may often be translated as impossible; time alone can enable you to decide in many of these cases whether the disease is malignant or not.   The chief points on which to rely are the age of the woman and the history of the case and the absence of tenderness.   Upon these particulars I shall not further enter, only insisting upon the great difficulty that exists in diagnosing pelvic cancer from chronic perimetritis, especially in the case of a young woman. And the difficulty is enhanced by the fact that even in old women perimetritis of all kinds, including perimetric abscess, may complicate pelvic cancerous disease.

Before I pass from the subject of pelvic cancer, I must mention another case accompanied by rather a rare symptom, discharge of fæces through the urethra.

S. N., aged thirty-six, married, has had two children and six miscarriages. The last child was born fourteen years ago. Was admitted March 8th, 1878, complaining of pain in left groin, which had lasted for fourteen years, but has been much aggravated the last five months. Micturition is frequent and scanty, and with the urine come occasionally air and fæces. The brim of the pelvis is occupied by dense hardness, not tender. On the right side an extension of hardness along the ischial plane and below the cervix, which itself presents no great abnormality. The uterus is fixed in this hardness. Its cavity is of natural length and direction.

This is a case which, if the hardness had not the long promontory coming down along the ischial plane, and other characters which are easily observed, but very difficult to describe verbally, would have been extremely difficult to diagnose from chronic perimetritis, because the woman was not elderly, and recently childbearing. The diagnosis was corroborated by the passage of air and fæces through her urethra. The passage of fæces per urethram is a rare occurrence, except in cases of malignant disease of the bowel, and especially the upper part of the rectum and the sigmoid flexure. You are not to suppose that the passage of fæces through the bladder is always the cause of much suffering, yet you would naturally think so. It generally only causes moderate suffering; in some cases, as in this, no suffering is mentioned at all. The passage of fæces through the bladder sometimes occurs in connection with peri- or parametric abscess, ending in intestino-vesical fistula. I have several times seen cases of chronic perimetric abscess where the abscess bursts into the bowel and also into the bladder. Such cases are diagnosed by their history. The fistula in such a case I have known spontaneously healed. Let me caution you against a supposition which I have more than once found prevalent in the minds of practitioners of otherwise great experience—that the passage of fæces through the bladder must of itself be fatal. Nothing of the sort. I have known patients with this infirmity live long lives, and die of other diseases quite unconnected with the passage of fæces through the bladder. It is, however, a rare occurrence, and always, on account of the rarity of its connection with anything else, suggests the idea of malignant disease. In the case I have just read, the existence of malignant disease was placed beyond doubt by the circumstances mentioned in the history of the case.

Now I come to a class of cases about which our knowledge is still very imperfect, and which, of late years, is getting more and more isolated from the general run, from those that would be called of uncertain seat,—cases of cancer of the body of the uterus. This is easily defined. A case is said to be of this kind if you have noticed it sufficiently early and find the body of the uterus affected by the cancer, while the neck of the uterus, so far as it is accessible, is healthy. It is a disease the rarity of which is exaggerated. Among the thirty-nine cases that I have mentioned, at least four were cases of malignant disease of the body of the uterus. This disease occurs in a variety of forms. I show you here, first, a magnificent specimen, an extremely rare one, of a uterus presenting diffuse, non-deforming, cancerous hypertrophy of the body of the uterus, the neck remaining, so far as the eye unaided, and the finger, can make out, quite healthy: a rare form of an uncommon disease.

The patient, an aged woman, began to suffer pain and think herself ill only about three months before she died. Her complaints were occasional attacks of pain in the hypogastrium, and occasional losses of blood per vaginam. She looked healthy for her years. Three weeks before her death she was admitted into the hospital under my care. A mobile hard tumor, of the size of a fœtal head, was felt projecting through the brim of the pelvis into the hypogastrium. It was rounded and not tender. She was seized with ordinary acute suppurative peritonitis, and sank in a few days. Cancerous nodules were found in the lungs and liver. The uterus weighed four pounds and a half, measured eight inches in length, and six inches and a half in breadth. Its cavity, from os tincæ to fundus, measured six inches. The walls of the body were about an inch thick. Examined by a competent histologist, the structure was declared to be that of hard cancer. Its section resembled that of a scirrhous mamma. The lining membrane of the body was thick and villous, only in some parts destroyed. There was cancerous degeneration of the ovaries; and a similar state of some limited parts of the vagina was discovered after death. The cervix, although healthy to appearance and to dig-

ital examination, was discovered by the microscope to be the subject
of cancerous degeneration. This case was diagnosed as a case of fibrous
tumor of the uterus; and, were it occurring in my practice again, I
have very little doubt I should again make the same mistake.

There are mistakes in medicine of which a man is ashamed; there
are others which do not make him blush in the least degree—and
this is one of them.   I do not know how I could make that diagnosis correctly.   The risk of error here is not like that in the diagnosis
of a case of cancer of the pelvis; you would never confuse diffuse
cancer of the body with perimetritis.   The diagnosis is between it
and fibrous tumor of the uterus.   If you look into books you will see
it justly remarked that one of the points of distinction is that in a
case of cancer the womb is fixed, and so it is generally; in this case
the womb was quite mobile.   Here, also, another usual symptom was
absent—there was no fetid discharge.   There was bleeding, but that
is also a symptom of uterine fibroid.   In this case the cavity of the
uterus was considerably lengthened, and so it often is in a uterine
fibroid.   In this case there were fits of pain, and these are not uncommon in a uterine fibroid.   You are led to suspect that a case is malignant—and at a first visit it is only suspicion—by regarding the history
of the case, the age of the woman (and I may remark that the age of
the woman is in cases of cancer of the body of the uterus greater than
in cases of cancer of the neck), the presence of an ascitic fluid in the
abdomen, and the induration and fixation which can sometimes be
made out of neighboring parts, especially of glands.   Of especial importance is the age at which the tumor began to grow, for a fibroid does
not begin to grow after the menopause. I have done enough to show
you how very difficult diagnosis may be in a case of this kind.

I have spoken of elongation of the cavity of the uterus, and it is
necessary to inculcate special care in making this out, in catechizing
the uterus, as it is often called.   In all cases of cancer of the uterus
is this care demanded, for then the uterus may be easily transfixed or
perforated by the probe; and this is not the case with an ordinary or
inflamed uterus.   Besides, the transfixion involves little or no danger
in ordinary cases.   I have known it frequently done, in the same
case even, without any evil result.   Yet it is always a misadventure
to be shunned.   The peritoneal wound does not gape or bleed in
such cases.   It is otherwise in examples of cancer of the body of
the uterus, and I have seen the fresh specimen in a case where this

gaping wound by the sound proved fatal within a few hours after its production.

Now, a few words on the mode of death. A woman with a uterine fibroid is not very rarely affected with chronic peritonitis of various kinds, sometimes causing a collection of peritoneal fluid to occur around it ; but this is very much more common in a case of malignant disease of the body of the uterus. In the present case you have another form of peritonitis exemplifying one of the modes of death in cases of cancer that is not very frequently described. Acute peritonitis of all kinds and chronic peritonitis are common with uterine cancer—local peritonitis, general peritonitis, and (the worst of all kinds) the acute suppurative peritonitis, which killed this woman in three days.

Besides peritonitis there are many other forms of death in cancer. It is only a specious concealment of ignorance that leads us to speak, as we often do in cases of cancer, of patients dying from exhaustion. I am very doubtful of that. No patient dies of exhaustion. You may say, "If a patient dies from bleeding, does she not die from exhaustion?" Very well; but that is dying from bleeding—that is, not undefined exhaustion. In the same way you find it often stated that patients die of pain. I never saw anybody die of pain, and I do not believe it occurs. So cases of cancer are said to end in death by exhaustion, as a man is said to die of old age. The truth, barely stated, is that you do not know of what he died. Now, the chief causes of death in cancer are peritonitis, urinæmia, septicæmia, pyæmia, and complications from diseases of veins or degenerations of important viscera. Sapræmia, or putrid poisoning (without addition of a living ferment), often causes fever and purging, and may cause even death.

The second form of cancer of the body of the uterus to which I will direct your attention is the nodular—a disease which makes the uterus resemble not a single uterine fibroid, but a group of uterine fibroids ; the nodules being different masses of malignant disease, deforming the uterus, almost certainly in this form of the disease fixing the uterus, almost certainly projecting into its interior, frequently bursting through and giving rise to bleeding and other fetid discharge, rarely bursting into the peritoneum and giving rise to fatal peritonitis. The second form of cancer of the uterus is not so rare as the former ; and here is a case of it.

M. L., aged fifty-nine, has been married for twenty-three years, and has never been pregnant. Complains of frequent and difficult micturition. Has constant pain in the lower part of the back and in the thighs. Has also a lump in the belly, which she says is increasing in size and has been felt for fifteen months. Her pains are severe at night, and she is rapidly losing flesh. Was in July an outpatient, and then had profuse fetid discharge, which has now ceased. Admitted February 22d. A layer of ascitic fluid intervenes between the abdominal wall and the tumor in the hypogastrium. The tumor projects most between the navel and the right spina ilii. It is hard and forms part of a large mass, which, projecting from the brim of the pelvis, extends to the left side of the hypogastric region. It is only sensitive, not tender. The cervix uteri, not notably altered, is high up and far back in the pelvis, and forms part of a solid hardness, fixed, and occupying the upper part of the pelvic excavation, and easily identified with the tumor felt in the hypogastrium. The bladder is not tender, but contracted, measuring three inches from orifice of urethra to fundus.

This example was easy of diagnosis only because the woman was fifty-nine years of age, at which time you do not get fibrous tumors growing rapidly with much pain as in this woman. There was, for this reason, no difficulty in diagnosing this case. There might have been great difficulty had she been a younger woman, and had we seen the case earlier. Then we should probably have had to watch it for a considerable time, for months, in order to satisfy ourselves as to its nature; but in an old woman, to have a tumor growing rapidly, fixing the uterus, pain always aggravated at night, ascitic fluid in the belly, forms a combination of clear indications.

I come now to other forms of cancer of the body of the womb, cancer of the interior of the body of the womb. I have just mentioned to you cases of ordinary (medullary) cancer of the uterus projecting into its cavity. When this happens—and indeed in all cases of cancer of the body of the uterus—you have to keep in view the distinction recently made (but not proved to be, clinically speaking, well founded) between the fibrous and the epithelial cancers, between sarcoma and common cancer. A sarcoma of the uterus has nearly the same clinical history as ordinary malignant disease such as I have been describing. Sarcoma is a malignant disease, only its progress seems to be generally a little slower than that of the ordinary forms of cancer, and it seems to be in a slight degree more amenable to treatment by removal. But really this distinction of cancers is too recent to have been fully followed out in its practical details.

The great malignant disease of the cavity of the body of the uterus is adenoma, a malignant glandular growth of the mucous membrane. Cases of this kind are not common. The growth bleeds, it distends the cavity of the uterus, fills it up, passes through the cervix, grows into the vagina, and I have seen a case where this malignant adenoma filled the vagina, and before the young woman's death protruded at the orifice of the vulva, the whole mass being composed of soft adenomatous tissue. In " Martha" we have had a case probably of this kind. It was sent in as an ordinary polypus, but on examining it, superficially even, it was observed to be very soft and fragile. The stalk went right through the cervix into the body of the uterus, and it was made out at the time of operation to be a case of polypus of the body of the uterus, not a fibrous polypus nor a mere mucous outgrowth or vegetation. On microscopical examination it was found to have all the structure of an adenoma. Dr. S. West found in it not only the uterine glands hypertrophied, and constituting the greater part of its bulk, but he also found in the centre of the tumor some muscular tissue; and a like observation has been made in some ordinary mucous polypi. Of this adenoma we have had no example except the polypus I have been describing.

I have lately seen cases of common malignant ulcer beginning high up in the uterine body, and such ulcers are quite different from the peculiar disease to be described in next paragraph. To get the diagnosis of such a case in its early stage, while the womb is mobile, it is necessary to dilate the cervix and pull the womb by volsella in its neck down upon the finger passed through it. Such cancers soon become more diffused, causing tumors and fixation of uterus.

The last malignant disease of the body of the uterus I have to mention is one affecting its cavity—namely, ulceration. The ulceration seems often to follow a previous condition of villosity. The villosity is destroyed, and ulceration takes its place; or ulceration is itself the commencement. This ulceration affects, like all malignant diseases, chiefly the old; and it has, in the vast majority of cases, the history of a malignant ulceration. But some recent investigations throw doubt upon the exact nature of the disease, although they do not entirely remove the malignant character from its ordinary clinical history. I am convinced that, speaking merely clinically, this dis-

ease in old women may be cured, if it is attended to early, by cauterization, by solution of nitrate of silver, of the inner surface of the
body of the uterus.   I have cured several cases of this kind where
there was copious discharge which was fetid, and copious bleeding;
and in some of which I have felt the seat of the disease with my finger, quite easily distinguished from the healthy surface of the uterus.
This feeling the seat of the disease has only been done after dilating
the neck of the womb by tangle tent.   In such cases, of course, the
disease is not—as yet, at least—malignant; and I shall say no more
of them.   In the more severe cases you may try the same treatment;
but when they get into this class they are irremediable.   The treatment may check the discharge, and produce great temporary improvement of health.   The patients die as in cases of ordinary cancer, sometimes with great suffering, and sometimes with little or none; and
after death, examination, as I have just said, leaves considerable doubt
as to the cancerous character of the disease.   In several cases that I
have examined lately there was no disease found except the ulceration
of the interior of the uterus, and that not of distinctly cancerous character.   In one which occurred in "Martha" there was found no evidence of real cancerous disease.   In that case the lumbar glands were
somewhat enlarged; but in other two cases even this evidence of extension was absent.

Ulceration of the cavity of the body of the uterus is characterized
by great pain in some cases, moderate pain in others, and in still
others no pain at all.   The pain is in some cases evidently spasmodic,
being so described and as resembling the pain of dysmenorrhœa, lasting only a few hours and returning daily or oftener, but occasionally
intermitting.   There are bleedings, which are sometimes slight and
sometimes severe.   The discharge is always very copious, not always
fetid, and may be purulent or ichorous.   The uterus enlarges, and,
instead of having little more than a potential cavity, may come to
have a cavity as large as would contain an orange.   The ulceration
extends deep into the tissue of the womb and destroys it; it comes to
affect the interior of the cervix, leaving the infra-vaginal portion un-
touched.   It sometimes goes on to perforate the peritoneum, and in this
way it may prove rapidly fatal; but I have seen, in one case lately, the
perforation met by adhesions, so that there was a peritoneal cavity or
abscess connected by a fistula with the interior of the uterus.   These
peritoneal cavities get filled, of course, with the same filthy discharge

which fills the interior of the uterus. The disease is easy of diagnosis. If you think proper you may go the length of dilating the cervix, so as to pass your finger in to feel the interior, and you may dilate the cervix for purposes of treatment—to wash out the interior of the uterus, and to cauterize it, if you think proper, with nitrate of silver or tincture of iodine. In all the cases which I have seen the disease has run a more or less rapid course, ending in death.

XVI.

UTERINE HÆMATOCELE.

THE subject of this lecture is uterine hæmatocele. Lately addressing you I mentioned an important variety of hæmatocele, which I told you Nélaton described as retro-uterine, and this description forms an extremely important event in the history of the disease. The case about which I am to speak to-day is not a typical one of retro-uterine hæmatocele. I wish it were, for the retro-uterine is an extremely characteristic species of the genus hæmatocele, and is easily described.

What is a hæmatocele? It cannot be defined in a few words. It is a tumor composed of blood, which may be in various conditions, such conditions being regulated chiefly by the age of the effusion. Do not imagine that every blood-swelling in or near the pelvis is a hæmatocele. Far from it. If the uterus itself, or an ovarian cyst, be distended with blood, that is not a hæmatocele.

In order to be a hæmatocele the blood must be *inclosed*. I prefer this term to *encysted*, which is that commonly used. It is objectionable because it conveys the idea of the existence of a special cyst, which there is not. Now, what is it that incloses the blood? The site of the effusion in the great majority of large and grave hæmatoceles is within the peritoneum, among the intestines, by which, and by parietal peritoneum, it is inclosed, the inclosure being completed by such adhesions as are necessary to make what may be called a cyst to hold it.

It is very common to say that hæmatoceles exist in the cellular tissue. Most assuredly they do, but I doubt whether these are the more numerous; at all events they are the less important. In these cases the inclosure is the cellular tissue, and the various organs. These I would much rather call by other names—hæmatoma, thrombus, or ecchymosis. Formerly, all hæmatoceles were thought to be in the cellular tissue ; but, chiefly through Bernutz, we have become enlightened on this point.

Now, suppose blood escapes into the peritoneum, it is not yet a hæmatocele ; but in time adhesions arise, which complete or fix the inclosing of the blood, and make it so.   For example, a woman has a tubal pregnancy ; about the third month the tube bursts, a large amount of blood escapes into the peritoneum, causing death.   This is not a hæmatocele.   Had the woman lived it would probably have become one.   I had an opportunity of examining a case of this kind before the blood became inclosed, and it is very difficult to diagnose such during life, because the effusion is so soft and so displaceable. When it becomes inclosed there is a recognizable tumor, and generally what would be called a hard one.   This is now the disease we have to speak of to-day.

Where does the blood come from ?   It is very difficult to say.   It may occur from the bursting of an extra-uterine pregnancy ; it has been verified as coming from the opening of a vein in the pampiniform plexus of the ovary.   A little phlebolite leads to ulceration, which gives rise to a small opening, through which blood is poured forth.   Rupture of the ovary has been proved to be a cause, not only the physiological rupture of a Graafian vesicle, but a rent in the whole tissues of the ovary.   In all such cases it is evident that the blood must escape into the peritoneum, and not into the cellular tissue. In the majority of cases the blood comes from that source whence, in a woman, bleedings are most frequent and important.   In menorrhagia, polypus, fibrous tumor, hæmorrhage following abortion, or delivery at full term, it is the mucous membrane lining the cavity of the uterus which bleeds; so it is, I am convinced, in the majority of uterine hæmatoceles, the blood flowing into the peritoneal cavity through a Fallopian tube.   The inner orifice of a tube is generally looked upon as always closed, and it is rarely seen otherwise ; but it is a sphincteric opening, like the cervix uteri, and is often felt by probe to be open.

Now for a few details regarding the case before us, which, though not retro-uterine, offers many valuable points for teaching.

A. B., aged twenty-three, married two years, never pregnant, began menstruation at fourteen years of age ; always regular, sometimes losing rather profusely.   About three weeks before entering the hospital the last period commenced ; continued for only three days instead of four, as usual.   Notice that a period stopped before the expected time. About two days after the cessation she was suddenly seized at night with pains in the abdomen, and in the morning she found a swelling in the lower part of the belly, which has remained ever since. Next day menstruation recommenced. Mark this also. She became feverish. When admitted the temperature was 100°, and she had a florid cheek. Note this expression; it was not a florid complexion (the term employed by the clinical clerk); but only a red spot on the cheek, the remainder being anæmic.

A large prominent swelling occupied the whole of the lower half of the belly, extending up to within two inches of the umbilicus; tender, dull on percussion, a smooth uniform surface, elastic, fixed. It became less and less tender, smaller and smaller, and about a fortnight after she entered the ward it had almost entirely disappeared. The temperature and pulse became natural. Nothing was to be felt but a little hardness, due to a few remaining adhesions. She declared herself quite well, and had then a florid complexion. The red spot on the cheek was lost.

Now, had this been a retro-uterine hæmatocele, the uterus would have been jammed against the pubes, and behind it you would have felt a large mass, like a retroverted gravid uterus. Instead of this, all that could be felt per vaginam by pressing high up behind the uterus, which was nearly in its natural position, was the lower part of the tumor, round, smooth, and tender. The case which I have described shows what a definite, well-marked disease uterine hæmatocele is. Nothing hazy about it. And when, later on, I speak of its history, you will be astonished.

I shall now tell you how I diagnosed this case. Unless the history is very distinct, well marked, and nearly sure, the diagnosis is very difficult. I have had many occasions to say—This is a hæmatocele —either before a woman's death or before opening the tumor, and when I have once decidedly said so I have not been wrong. What are the points which have guided me to a conclusion? They are all well illustrated in this case.

First of all I must tell you that what we generally have to diagnose it from is an abscess, retro-uterine or other, and there is frequently great difficulty. There are three principal points: 1st, Suddenness of symptoms, and suddenness of tumor. 2dly, Derangement of the menstrual function, in the shape of arrest; in the most typical cases there is a sudden stoppage, and then the pain. 3dly, Anæmia.

When this woman came into the ward, the history not having been taken, I diagnosed hæmatocele, but with only a very moderate degree of assurance. The diagnosis was from an abscess. In a former lecture I described a case of retro-uterine perimetric abscess, which was very like this in its physical characters, but there was the absence of suddenness or menstrual arrest. The symptoms of suppurative fever were also present.

My diagnosis was a direct one, and the induction was from the three circumstances which I have enumerated. I assumed that we should probably have another element of direct diagnosis in the future history of the case. When the disease had lasted five weeks—that is, a fortnight after admission to the ward—it was all gone, and nothing had come out of it. There had been no evacuation of pus, no diarrhœa. The patient had been getting better every day, and the lump was melting away like a snowball in the sun.

I will take this opportunity to say a few words about diagnosis. It is the first step in all medicine, an art which is in a very uncertain state. I have often seen practitioners prescribe, and then set their brains to work to ascertain what is the matter. When you treat a disease without knowing what it is, it is like shooting crows with your eyes shut. Diagnosis is, therefore, what we must first aim at, direct diagnosis if possible. But we are often glad of another—a limping method—diagnosis by exclusion, a method founded on the axiom—If we don't suspect a disease, we shall not be likely to find it out.

Here is an abdominal tumor, and we may commence to exclude. A cyst is suggested—an ovarian cyst, a parovarian cyst; a renal tumor, or hydronephrosis; a perimetric abscess; hydatids; then, perhaps, a fibrous tumor, or pregnancy. We run over these hastily in our minds. Can it be one of them? It is a very shabby method of diagnosis, but we are bound to use it in order to do our best for the patient. The history of this case proved that the diagnosis, made both directly and by exclusion, was correct. There is no tumor that I know of that will go away thus rapidly while a woman is lying in bed, without any evident evacuation, except one composed of blood. The diagnosis was not only direct and by the history, but also by exclusion. Our last case of perimetric abscess, a tumor like this hæmatocele, went away quite as quickly, but pus flowed in torrents from her bladder.

Now, from what you have heard of the case before us, you might say this is a very trifling disease—cured at once. It would, however, be a very wrong idea. It is not every such hæmatocele that goes on like this. Sometimes the tumor increases instead of diminishing; or it diminishes and then suddenly increases again; or the peritonitis which is induced may not be simple adhesive, but a great abscess may form; or the blood may putrefy and produce septicæmia. Or the tumor may burst into the bowel, which, though often a fortunate termination, occasionally leads to septicæmia by feculent matters getting into the cyst. Sometimes the tumor will not go away, absorption will not take place; why, I know not; but it may have something to do with pressure relations.

With regard to the treatment. The patient was simply kept in bed, and this is the most important treatment of all. Probably, had she moved about, the result would have been very different. But sometimes we have to direct our attention to endeavor to stop the bleeding. I have no great confidence in anything for this, but I will tell you those remedies which appear the best, and in their order of merit—1st, perfect rest; 2d, ergot of rye; 3d, ice poultices; I fancy I have seen benefit from these last, but I have also, I believe, seen harm. Then, if you have been brought up in the antiquated school, you will believe in sorbefacients; I don't believe in them. Muriate of ammonia lotion; tincture of iodine. You may prescribe them if you like, for perhaps they will please the patient. The further treatment depends upon circumstances.

It is sometimes very difficult to decide whether or not to evacuate the cyst. In the great majority of cases it is unnecessary. I have used both knife and trocar, and I do not see the objections to their employment which are entertained by most gynæcologists.

After opening the cyst I advise you to take care of two things. In the book of an eminent gynæcologist you are recommended to insert the finger through the artificial opening, and break down bands or adhesions in order to let free the blood mass. The writer was under the delusion that the blood is situated in the cellular tissue; it is, however, in the peritoneum, and the adhesions are the safety of the woman. If, therefore, you attempt to break these down you will be doing the very worst thing.

Another treatment which has led to many fatal results, is injecting the cavity by means of a strong syringe, the consequence of which is that the beautiful protective arrangement of nature is damaged. I have seen many women with a fetid discharge, and all the intensest symptoms of sapræmia or putrid intoxication; and as soon as the cause was got rid of they got well.

I now come to what I told you would astonish you—the history of the disease.

It was unknown when I was a student. I studied medicine in Aberdeen, Edinburgh, London, and Paris, and in none of these places did I hear of such a disease. It is now most difficult to conceive how this most manifest disease could have passed unnoticed. If you consider a gold field—at first great nuggets are discovered, then smaller ones; then to find gold the sand has to be sifted; and lastly, there is none at all. So in anatomy and pathology, great discoveries were at one time easily made, but now we have to get a microscope with a lens of the highest power to find anything new. When I was a student we had only morbid anatomy; now we hear of nothing but pathological histology.

With regard to uterine hæmatocele. Here was a nugget of the largest size which remained practically undiscovered till a few years ago! This is a remarkable subject for reflection, and shows us how carefully we should scrutinize our cases, for there may be some as great nuggets buried in the field of medicine even now, when we think the time for such gross discoveries is past.

These great tumors must have existed in former times. What did physicians make of them? I have read of the cure of large fibrous tumors in a week or two, of large ovarian cysts being dispersed in a very short time by some marvellous medicine. You will find among good authors plenty of such cures. No doubt some of these so-called tumors were hæmatoceles, and would have disappeared equally quickly without the imposing remedies.

XVII.

PAROVARIAN DROPSY.

THE subject of this lecture is a case of simple parovarian dropsy, which has just been dismissed from "Martha." It is not, in respect of the fluid drawn off, a perfectly characteristic example of the disease, but it is on the whole very nearly so, and well worthy of your attention.

In this region of the body there occur several kinds of cysts, besides the well-known fibro-cystic uterine disease, simple ovarian cysts, dermoid cysts, and the different kinds of multilocular dropsy of the ovary. There are the cysts sometimes named after Follin, little blebs, which are frequently very numerous, scattered over the tubes and broad ligaments; they are not discoverable during life. There are the metro-peritonitic cysts of Huguier, perhaps rather a kind of vesicular œdema than a true cystic formation; these also have, as yet, no practical significance. There may be cysts of the ducts of Gärtner. There is often observed in post-mortem examinations a little cyst hanging by a long stalk from the outer end of a Fallopian tube; it corresponds with the hydatid of Morgagni in the testicle, and is the dilated closed end of the duct of Müller, the part of which nearer the uterus is transformed into the tube and its infundibulum.

The parovarium in the female corresponds to the epididymis in the male; and it consists of a series of tubules running from the hilus of the ovary along its mesentery towards the neighboring tube. In the disease, of which we have had recently a fine specimen in "Martha," one of these tubules is dropsical, distended with a thin fluid. It is said that the affected tubule is generally one of those most distant from the uterus. In most cases, but not invariably, one tubule only is affected, and the cyst is truly, anatomically, unilocular. It has been observed to be bilocular in rare examples, or even trilocular; but, in the disease we are now describing, the cyst is never multilocular, never proliferating.

Some authorities say that cysts, indistinguishable from the parovarian cyst during life, do occur in the ovary—simple serous ovarian cysts. I have seen many such, but never one that was of great practical importance from its size, never one that might be confounded with multilocular ovarian disease. The disease we are now considering is simple parovarian cyst. We are not at present concerned with complicated cases, or cases not simple, whether the complication be inflammation of the cyst, hæmorrhage into it, or the occurrence of proliferation in a malignant form, such as Olshausen has recently described.

Simple parovarian dropsy is an important disease, and very alarming, for it naturally excites suspicion of the presence of the terrible ovarian cystoma, or multilocular ovarian cyst. Twenty years ago, or even less, it was generally confounded with this disease, and there can be no doubt that many of the spontaneous or artificial cures of ovarian dropsy were not so, but really cures of this disease— the simple parovarian cyst. It is spontaneously cured by rupture, during pregnancy, or apart from that state. It may, perhaps, be cured by absorption of the fluid. It is often cured by one tapping. But the present state of our knowledge indicates that an ordinary ovarian cystoma is never spontaneously cured, never entirely disappears. Ovarian cystoma is generally cured only by ovariotomy. Yet there is no doubt that, in some rare cases, an ovarian cystoma gets smaller, its contents partially absorbed or inspissated; it, indeed, is sometimes spontaneously virtually cured, but very rarely.

Next to the truly unilocular condition comes, as an important feature of this disease, the character of the fluid. It is almost pure water, having a peculiar opalescence like that of lime-water, or of a quinine solution. Very little or no albumen is found in it, but appropriate tests show the presence of the chlorides of sodium and potassium. In the sediment there may be occasionally detected cylindrical epithelial cells in the midst of other detritus. The specific gravity of the fluid is low, generally, as in our case, under 1008; whereas that of ordinary ovarian dropsy is much higher, ranging from 1010 to 1025, or still more. In our case the fluid was not perfectly characteristic, but nearly so; it had a yellowish tinge, probably from some slight mixture of blood with the fluid taking place a long time ago. It did not otherwise vary from the regular parovarian fluid. In a case which is not simple the fluid may have quite other characters, from the admixture of pus or of blood, or of both in various quantities. I have seen it like honey in consistence, and like coffee grounds in appearance.

A parovarian cyst was, till recently, supposed never to attain a considerable size, seldom to be larger than a fœtal head; and this was very misleading, for the dimensions may be enormous. Here I show you one which is far larger than a gravid uterus at full term; it would, indeed, easily accommodate several adult fœtuses.

The characters of a simple parovarian cyst which I have gone over are to be made out during life. After death, or after the removal of the cyst, you find other distinctive characters. The more important are the great elongation of the tube around the cyst, at least at its external part, or the part remote from the uterus. Another of the more important characters is the easy peeling off of the peritoneal coat, or enucleation of the cyst proper from its peritoneal investing sac. In the case of an ovarian cystoma there is no peritoneal coat, and, if you try to tear off an outer albugineous coat, you merely strip off irregular patches, producing nothing like the easy separation of coats seen in the true simple parovarian cyst.

Such is a sketch of the disease we have illustrated in the case of M. M., who has recently left the hospital—a disease which in her has been at a standstill for about three years, and has now at length, on account of its cumbersomeness, led her to seek its removal.

The chief facts of the case are as follows: M. M., aged thirty-nine, married, has had seven children and three miscarriages. Her last pregnancy ended naturally five years ago, the delivery being completed by forceps. The catamenia began at seventeen years of age, and have been generally regular. Three weeks after her last confinement she observed that her abdomen was of the same size as before her delivery, and for two years it continued to increase. Since then she thinks it has been stationary. The abdomen is very large, semi-globose, distended from pubes to sternum. Over its anterior surface and well backwards towards the flanks there is absolute dulness on percussion. Over every part of the dulness, and in every direction, there is perfect fluctuation. The most prominent part of the belly is three inches above the umbilicus, and here the circumference is forty-two inches. At the umbilicus it is forty-one. The distance from the ensiform cartilage to the umbilicus is ten inches; from the umbilicus to the pubes seven and a half inches; from the umbilicus to the right anterior spine eleven inches; to the left ten and a half.

There is no fever or derangement of any kind.

You will observe,—not a word of complaint. In truth, the enlargement and the weight of 400 ounces of water produced no symptoms proper; and it was plain the poor woman scarcely thought it worth while to have anything done for herself. She felt no need of relief. This absence of symptoms is a very important matter, for it shows that this disease has no essential or necessary symptoms; and the same is true of ovarian cystoma. Many diseases have essential symptoms, of which the most common is pain. Here we have none. Every case is not without symptoms, even varied kinds of suffering. But the utter absence of them in our case shows that their absence is no indication of absence of disease.

While there were no symptoms, the signs of disease were very distinct, and in a great degree distinctive. The short statement of the chief phenomena of this case that I have already given describes the signs. These signs enable us to diagnose the nature of the case. The direct diagnosis is, however, not so perfect as to enable us to dispense with the differential diagnosis or diagnosis by exclusion, but it is nearly so. It is only the direct diagnosis that I shall have time to make any remarks on to-day. The direct diagnosis enables us with considerable assurance to say—this is a case of simple parovarian cyst. The differential diagnosis justifies us in saying—this is not hydramnios, not ascites, not chronic peritonitis with effusion, not ovarian dropsy, not fibro-cystic disease of the uterus, etc.

The abdomen was greatly enlarged, and had a smooth hemispheroidal outline with no irregularities; it felt as if full of fluid. These circumstances are consistent with unilocularity. It had a projecting, rounded shaped, not loosely flattened form; there was no history of disease that might produce it, no evidence of peritoneal adhesions around it, no change of the area of resonance on changing the position of the patient—circumstances which indicate that the fluid is encysted. The repletion of the cyst with a thin fluid was not made certain by its feeling as if full of fluid, but was made certain by perfect fluctuation producible everywhere in it. The perfection of the fluctuation in every direction, and the wave being easily produced from any part to every other part, showed that the cyst was unilocular. Thus we diagnose a unilocular cyst full of thin fluid. But this does not complete the diagnosis.

Before advancing I wish to impress on you some very important matters regarding "feeling fluid" and "fluctuation," terms which are generally misconceived and misapplied. The true appreciation of these valuable signs will save you from many and frequent errors.

Feeling fluid is a very common sign. It is often, indeed generally, called feeling fluctuation; but it is quite another thing. When in the midst of inflammatory induration you feel a soft fluid portion, you have a high degree of assurance of the presence of fluid; but this assurance comes as much from the history of the softened part as from the actual sign. The history and the sign together may in many instances give you a high degree of assurance, approaching to certainty. The feeling alone is very deceptive, and it is when alone that you have to study it in order to make out its value. I know few more prolific sources of error than confidence in "feeling fluid." Dry tapping, as it is called, does not always show an error in the operator; for he may have tapped while conscious of uncertainty as to the presence of fluid. But dry tapping is, after all, a common error. How often is an inflamed mamma incised when there is no abscess, but only the misleading feeling of fluid?

You should all carefully learn the invaluable sign "fluctuation" in a case like the one we are now describing. You percuss or gently strike with a finger or fingers, and produce a wave, which your other hand or the fingers of it receive. It has to be distinguished from a communicated impulse, which may be transmitted through soft parts which contain no fluid. When you feel fluctuation, you have a valuable positive sign of fluid—an infallible sign. You must not say you think you feel fluctuation; for then you had better say you do not feel it. You either feel it or not, just as you feel the pulse or not. If you feel it, you do not say you think there is fluid; you say there is fluid. These important points I have no time at present further to insist upon.

In the case before us, perfect fluctuation could be easily produced between any two parts of the cyst. This is another valuable sign. It shows that the cyst is unilocular—a single-chambered bag. Were there two or more large chambers, the fluctuation would not be perfect in every direction. The dissepiments between the chambers would arrest the wave more or less completely.

Let us now consider what we mean by unilocular. I have told you that our parovarian cyst is unilocular. It is truly, or anatomically, or scientifically, or absolutely unilocular. Again, I have told you that by use of the sign, fluctuation, we have diagnosed its unilocularity; but, in truth, we have not diagnosed its real or anatomical unilocular condition. We have only made out that it is surgically unilocular; unilocular for such purposes as those of the ovariotomist. Small cysts in the wall of the large cyst, or connected with it, would not damage the fluctuation sign of unilocularity. It is, therefore, only a surgical or conditional unilocularity that is shown by this sign.

We tapped this cyst in the ordinary way, and drew off twenty pints of fluid. After tapping, we had another evidence of the unilocular character of the cyst. We could feel no cyst at all. It had not collapsed into a ball or mass, as it sometimes does, but lay so as not to be felt. I have felt such cysts after they have shrunk; but even then not distinctly. After tapping, the bowels descended and filled the belly everywhere, resonance being produced on percussing every part. No coherent masses of bowels were felt; as is usual after tapping the fluid collected in a case of chronic peritonitis.

I may here mention to you a rare dissection recorded by Professor Gairdner, which shows what happened to a parovarian cyst in one case. His patient had a large abdominal swelling, produced by a great cyst, which suddenly and unexpectedly burst, and the swelling disappeared. Sixteen months thereafter she died of Bright's disease. Dissection revealed a small parovarian cyst, which was empty, and if distended might have equalled in size a fœtal head. The place of rupture was made out by Dr. Coats, and was shown me by Professor Gairdner. Though the rupture was healed, the cyst had shrunk and had not re-filled.

The case illustrates the treatment of the disease. When simple, as in M——'s case, it is often cured by one tapping. What becomes of the cyst we may guess from the state of it in Professor Gairdner's case. I have tapped several such cases, where I have for years followed the patient, and found the cure permanent.

But all parovarian cysts are, unfortunately, not susceptible of such easy and successful treatment. Complicated cases may even require an operation like ovariotomy. I have seen several such operations, where there was extreme difficulty from adhesions, and from the thinness and lacerability of the sac. The proper treatment of complicated cases is not yet decided.

In simple parovarian cysts your course is plain. By tapping and examining the fluid you complete your diagnosis, you relieve the patient's and your own mind from fear of disease of a graver kind, and you hold out to your patient the prospect of complete and permanent relief.

XVIII.

RUPTURE OF OVARIAN CYSTOMA.

Of this accident we have recently had an example in "Martha" Ward, and I shall commence the lecture by reading an account of it.

S. L., aged forty-nine; married nine years; one child eight years ago. Catamenia began at seventeen, and ceased two years ago. No definite history of the present illness can be obtained. She says that she has been confined to her bed for three months; has suffered for some time from constipation, and also from vomiting. Has not noticed any lump in her abdomen. Is emaciated.

When admitted, suffering from constant vomiting of a dark-green fluid. Countenance pinched and anxious. Pulse small and feeble, 132. Temperature 99.6.° Lies on her side; legs drawn up. Breath has smell of new-mown hay. Has frequent eructations. Belly very prominent and tight; measures at umbilicus $35\frac{1}{2}$ inches, is resonant in nearly every part, presents fluctuation beneath the umbilicus from side to side. Brim of pelvis occupied by great fulness and hardness, as felt per vaginam.

Ordered to be fed with iced-milk and beef-tea; to have hypodermic injections of morphia to allay pain; and to have a careful trial of the best means for subduing vomiting.

She was hopelessly ill—indeed, almost moribund on coming to the hospital; and she died four days after admission.

Post-mortem, Fifty-five Hours after Death.—Body somewhat wasted. Rigor mortis well marked. On opening the belly, air at once escaped, subsequently followed by a yellow puslike fluid. At the lower part of the belly was a large tumor, filling this part and the whole of the pelvis. Above the tumor was a cavity from which the greater part of the fluid escaped. This cavity had for walls the anterior part of the small intestines at the back; the omentum and abdominal walls at the front; above, the transverse colon. This cavity was clearly marked off by firm adhesions from the rest of the peritoneal cavity,

and was larger in size than a man's head. The small intestines were
to the left of the cyst. Liver, spleen, stomach, and intestines all
matted together by old adhesions. Stomach and intestines natural on
mucous surfaces. Liver pale, friable, fatty. Kidneys small; cap-
sules somewhat adherent. Cortex of natural breadth, but pale and
indistinct; pyramids pinkish.

On lifting up the cystoma at the bottom of the belly there was seen
in a cyst to the right, overhanging the linea innominata, a gaping
aperture the size of a florin, communicating with the cavity in the
peritoneum described above. The fluid flowing from a burst cyst was
dirty yellow, like a mixture of ovarian fluid and pus. Around the
orifice the tissue of the cyst was in a sloughing condition for a considerable distance from the margin. There were old adhesions, between the body of the uterus and the large mass of the cyst, springing
from the left ovary. The cysts were ordinary ovarian cysts, the
largest being about the size of a cocoanut, holding gumlike or
honeylike fluid; in some the fluid was thinner. Right ovary natural. Bladder somewhat injected. Uterus natural. Vagina mauve-
tinted.

Here was a remarkable and very interesting case. You observe
the symptoms and signs were, very quick pulse, temperature slightly
elevated, uncontrollable vomiting, and a tympanitic abdomen, no
tumor to be felt except by vaginal examination, and then only hardness in the brim of the pelvis, suggesting little more than the idea of
a tumor. With these signs, and an imperfect history, the diagnosis
was extremely difficult or insecure; and when I said, "I fancy this
is a case of burst cyst," it was more a conjecture than a diagnosis. It
turned out to be thus far true; but, in some other respects, my ideas
respecting the case were not altogether correct, for I believed her to
be dying from peritonitis acutissima, as a consequence of the bursting
of a cyst. What were the signs which led me to this last conclusion?
They were an extremely distended tympanitic abdomen, with intense
tenderness, and distressing uncontrollable vomiting. But, if you have
followed the account of the post-mortem examination, you will have
seen there was no acute general peritonitis, for there were extensive
old adhesions which had nothing to do with death, but which had an
important influence on the progress of the case, limiting the diffusion
of the irritated escaped ovarian fluid and the consequent peritonitis.
The ovarian cyst was lying in a peritoneal abscess, the parts around
being matted together by recent lymph, forming a great abscess cavity,

in the bottom of which lay the ovarian cyst, nearly the size of a man's head.   You observe she had an intra-peritoneal abscess as the result of this rupture.   The old adhesions saved her from acute diffuse suppurative peritonitis.

She did not die from peritonitis, nor from this peritoneal abscess, but from putrefaction of the sloughing cyst and its contents, which developed a quantity of gas in the sac of the abscess, and thus gave rise to a form of abdominal tympanitis.   Now this is not an ordinary condition, and the inquiry is suggested : Why should a slough which is under antiseptic conditions putrefy?   This is probably from the neighborhood of the bowels ; and there are many analogous cases.   I have seen the same result in a case of pelvic hæmatocele, when probably nothing had been effused into the peritoneum except pure blood ; but we must dismiss this subject.   Our patient died of septicæmia, her blood becoming poisoned by the absorption of putrid matter.   The intensely strong smell of newly mown hay from the breath was indicative of this, as well as the whole progress of the case.   The mere burst cyst and the intra-peritoneal abscess do not account for the case.   She might have made good her recovery had there not been septicæmia.

Ruptures occur frequently in women, in the lower abdomen, and especially during the childbearing period of life.   First, there is the periodical rupture of the Graafian follicle, from which escapes the ovum, the fluid of the vesicle, and possibly a little blood ; the rupture being sufficiently large to admit the extremity of a good-sized probe. This is a physiological rupture ; but pathological ruptures in this situation are not uncommon.   There occurs rupture of the ovary itself, a lesion which has occasionally led to fatal results ; the ovary being found split open as though it had been incised by a dissecting knife. Well-known ruptures occur in the uterus in connection with delivery. There happens also occasionally, during delivery, a curious rupture of the peritoneum covering the uterus, which is, as yet, inexplicable ; a beautiful example of this has been shown to me at this hospital by Mr. Butlin.   Then, during pregnancy, there has been observed erosion commencing in the peritoneum, and penetrating a large uterine sinus, resulting fatally by peritoneal hæmorrhage.   Also, a vein in the broad ligament may give way, just as a varicose vein in the limbs does.   This is said to be the result of ulceration produced by a phlebolite, and the consequence is escape of blood.   Another rupture is that of the Fallopian tube in extra-uterine pregnancy.   I am sure that the rupture of small serous cysts in the ovary is quite common. Sometimes a small ovarian cystoma, not larger than an orange, will rupture. I have myself seen an example of this. A woman was found lying dead in the road, supposed to have been ill-used; a judicial examination was made, and death was found to have been due to haemorrhage from the rupture of such a cyst. Then, frequently, small cysts situated on the surface of larger ones burst; as may often be observed when ovariotomy is performed.

What I have more particularly to speak of is the bursting of a large cyst, of which so good an example has formed the basis of this lecture. Rupture of such a cyst may be produced by ulceration, but this is probably rarely a cause, the influence being generally mechanical; either distension by blood, pus, or ovarian fluid, or external violence. It may occur from handling a cyst, without such rudeness as is called violence. In some cysts the walls are very thin, and they become softened by inflammation, or fatty degeneration, or sloughing, so as to be rendered excessively lacerable. This teaches us that ovarian cysts should always be very gently handled. I have seen a cyst so frail that slight pressure at one part made it burst at a remote part. The case just read was one of rupture by ulceration and sloughing; and we have lately had the sudden death by peritonitis of a woman who was waiting to be operated on by ovariotomy, and in her the rupture was by ulceration, a round hole of the size of a sixpence being found after death, looking as if it had been punched out.

If the cyst bursts, what results? The woman may bleed to death, if bleeding into the cyst be the cause of rupture. Rupture of a cyst probably always produces a certain amount of peritonitis, generally of a kind of which little is known. If a cyst burst, the fluid of which is quite bland, often no pain is produced, but probably this low form of peritonitis which may last a long time without even producing adhesions. The peritoneum is red and raw-looking, sometimes with large areas which appear covered with a granular lymphy deposit. This peritonitis produces friction, which may be sensible to the hand and ear; and the granular condition may even be felt when the abdominal wall is very thin.

But if the fluid be mixed with pus or with old grumous blood, then probably acute peritonitis will arise and rapidly supervening death.

The case before us illustrates another danger, that of septicæmia. It occurred in connection with a vast peritoneal abscess in this case;

but septicæmia may be caused in another way. If the cyst burst into the bowel, generally relief follows; but occasionally, from regurgitation and extravasation of feculent matters into the cyst, chronic septicæmia is set up, the woman dying very slowly. I know of cases, well recorded, where women have lived for years after an accident of such a kind, where at least gases from the bowel entered the cyst. I have never myself seen such a long survival, but ordinary air is frequently admitted by misadventure into cysts during tapping, without any harm accruing.

The results of burst cyst depend greatly on the character of the fluid effused. If it be pure blood it will not necessarily excite acute general peritonitis; if it be thin and bland ovarian fluid it will do little harm. It is alleged that, if it be very thick and viscid, it will cause acute peritonitis; but from several examples, verified by post-mortem examination or observed during ovariotomy, I can say that this is, at least often, not the case. If, however, the fluid be pus or grumous blood, or contain them, for a certainty acute peritonitis will arise, and speedy death follow unless ovariotomy be at once performed.

The escaped fluid generally passes freely among the bowels, but not always; for its progress may be restrained by old adhesions as in our case; or it may be so viscid as not to become diffused in the abdominal cavity, but displace the bowels as if it were itself a tumor.

When the fluid is in the peritoneal cavity it may be easily diagnosable as free fluid, if it is in large quantity. Sometimes it would appear to become inspissated, the watery part only being absorbed. Viscid fluid, after extravasation into the peritoneal cavity, is often difficult of diagnosis as free fluid, for it may displace the bowels as a tumor does.

All I have said is on the supposition that the cyst bursts into the peritoneum. But not very rarely it ruptures into some of the mucous passages. Generally this is made plain by the escape of the ovarian fluid from the body, and by the diminution of the size of the cyst, but it may occur so insidiously as not to be discovered by the physician. Such ruptures into the bowel have never been healed, so far as I know. I have known and put on record a case of burst cyst, where the fluid was discharged per vaginam, and where, long afterwards, on ovariotomy being performed, no adhesion of the cystoma to the pelvic organs was discovered. The rupture must have healed.

The same healing occurs in many cases of a cyst bursting into the peritoneum, the cyst refilling and rebursting, sometimes repeatedly. But occasionally the aperture in the cyst remains widely open, as is sometimes finely seen when the fluid is viscid and distends the opening in the cyst while it is exposed to view, in ovariotomy or in a post-mortem examination.

Ovarian fluid may be quickly absorbed from the peritoneal cavity, and sometimes seems to be discharged from the system, by the kidneys or by the cutaneous surface. It is doubtful whether the peritoneum can absorb the very viscid fluids. Certainly in some cases viscid fluid lies in the abdomen for months or years, apparently slowly accumulating rather than disappearing by absorption.

The practical importance of rupture of an ovarian cystoma is very great. Bleeding into a cyst, no longer restrained by the resistance of the cyst-wall, may go on into the peritoneum to prove rapidly fatal. The escape of irritating fluids from a cyst may induce diffuse peritonitis, or, as in our case, extensive peritoneal abscess. Further, septicæmia may be produced by intra-peritoneal putrefaction, as is also exemplified in the case before us.

Diagnosis may be rendered difficult, if not impossible, especially if the history of the case be not fully known. With a full history the diagnosis will probably be easy, for then the fluid lying free in the abdomen will not likely be mistaken for ascitic fluid or for a collection of fluid in a case of chronic peritonitis, as otherwise it might well be.

You know that, in all operations, you are advised to go over the diagnosis once more just before you begin. In none is the value of this more frequently exemplified than in ovariotomy and ovarian tapping. I have been on the point of tapping a large ovarian cyst, when I discovered that it had ruptured, that there was no distended cyst, but an abdomen filled with the escaped fluid (which was subsequently rapidly absorbed). I have seen a case in which ovariotomy was just about to be done, in which it was unexpectedly found that the cyst was tympanitic, a large communication (as the autopsy too soon showed) having formed between the chief cyst and the great intestine at its sigmoid flexure.

Rupture of ovarian cystoma does not always prevent ovariotomy. Sometimes, indeed, as I have already said, it demands immediate interference. Sometimes it only leads to delay of the operation, as was the case in the instance of intended tapping which I have just noticed; and as in a case of bursting and evacuation of fluid through the vagina, which I have also mentioned in this lecture. Sometimes this bursting, when it takes place into the bowel, prevents ovariotomy altogether; at least I know no case in which ovariotomy has been successfully done, or even deliberately attempted, when a communication between bowel and cyst existed. This complication presents difficulties which the great ingenuity of our operators has not yet vanquished.

Finally, rupture of ovarian cystoma is an accident, the risk of which must be considered in deciding the very important question: When should ovariotomy be done? Few of you may ever become ovariotomists, but probably all of you will be called to assist in the decision of this important question, and you must not neglect this element in arriving at a conclusion. Some ovariotomists prefer operating on young, robust women at a comparatively early period of the disease, before the cyst gets very large. Other ovariotomists prefer operating on women when the cyst is comparatively old and large, and when their health, if not positively injured, is at least in a very degraded condition, when they are wasted and oppressed with the disease. The question is a very difficult one, to be decided by experience and according to the merits of each individual case. It is a question into which considerations of humanity as well as of surgery enter largely. You must not look upon your patients as mere cases. The mere case is only a part, an exclusively surgical part, of the whole. You must advise your patient, remembering, if not strictly obeying in every case, the grand precept to do to others as you would be done by. Now a consideration of humanity does, I know, powerfully affect ovariotomists. They are unwilling to subject a woman who may live happily for a year or even years to the risk of death within a few days from ovariotomy; while, in the case of a woman exhausted by the disease, whose life is not worth many days' or weeks' purchase, they have no such scruple. In the final decision of this important question of when to perform ovariotomy, the danger of rupture of ovarian cystoma must have a weighty part. Had we had means and opportunity of foreseeing the accident which happened to our immediate patient, we should not have hesitated to recommend early interference by excision of the tumor.

XIX.

PROCIDENTIA UTERI.

THE subject of this lecture is one of the most important among the diseases of women,—Procidentia of the Uterus. It is of the simplest kind, nearly purely mechanical, quite as much so as a dislocation of the shoulder, or a hernia. There are a variety of other views held which may be called vital, connecting it with some diseased condition as a cause, but I am satisfied that it is mainly mechanical. In the descent of the uterus there is a variety of degrees. The first is generally called descent; it is the slightest degree. The second is prolapsus, in which the neck of the womb is near the orifice of the vagina. The example before us is, however, in the most important degree, procidentia, a falling forth from the body.

When the patient came to us the womb was not procident, it was merely in a condition of prolapse, lying on the perinæum, not outside the woman's body. But if she walked about, or made any effort, it came outside; therefore it is classed among the cases of procidentia of the womb.

Now, what makes a woman's womb fall out of her body? To investigate this, we must inquire what keeps it in its place. The most important cause is the pressure relations of the abdomen. The womb floats. Suppose in the corpse of a healthy female you open the abdomen, the womb is then always found in a state of descent, because the destruction of the entirety of the abdomen robs it of its support. Before the abdomen was opened the uterus was in its normal position, the fundus about on a level with the brim of the pelvis.

If I were to ask a first year's student what keeps the womb in position, he would at once answer—the ligaments. The idea is, however, quite an erroneous one; the term *ligaments* as applied to the utero-sacral, the utero-vesical, the round, and the broad ligaments, is a most unfortunate one. They are not ligaments at all. If they were they would prevent the womb from moving, whereas their func-

tion is to give it unlimited motion. They stretch and give to any
extent, if a due amount of time is allowed. The next force which is
said to keep the uterus in place is the perinæum, and the state of this
is a very important matter in the case of procidentia. In a healthy
woman the labia are separated only by a line, and between their junction posteriorly and the anus is a considerable space, the perinæum
proper. In procidentia it is quite different; instead of the labia
majora meeting one another, there is a great gaping orifice, into which
you might put your fist; through this the womb is easily protruded.

It is generally and erroneously stated that rupture of the perinæum
during childbirth is a great cause of procidentia. There is a wide
difference between causing and facilitating an event. If you were to
take a healthy woman and put a knife in at her anus, and bring it
out at the fourchette, the womb would not alter its position. How
do we know this? Because Nature has demonstrated it by experiment. I have seen many cases where even the recto-vaginal septum
was torn through, and there has not in any of them occurred a prolapsus of the womb. Therefore rupture of the perinæum has nothing
whatever to do with causing procidentia. But it has to do with
facilitating it.

The birth of a child over the perinæum may be compared to the
birth of the womb over the perinæum. In childbirth the first thing
that occasions waiting is the opening of the mouth of the womb;
the next is the distension of the rigid perinæum. Take, however, a
woman who is not only a multipara, but whose perinæum has been
lacerated; as soon as the head of the child gets through the os uteri,
there is nothing to stop it. So it is with the womb when it reaches
a state of prolapsus; the perinæum having been torn there is nothing
to stop it; it is outside at once.

Procidentia is, therefore, more likely to occur to a multipara than
a primipara, but even a virgin may suffer from it, and I have seen it
before menstruation has commenced. Of the peculiarities of this
procidentia in early life I have no time to speak to-day.

In the case before us we had to consider the state of the perinæum.
It did not show much sign of laceration. But on examining the parts
I had occasion to comment upon a statement of the woman. She
asserted herself to be a virgin, and yet I am satisfied that she has had
at least one large child. I had no discussion with her, for whatever
she might say would not alter my opinion. And these are the rea-

12

sons for my decision: First, I found in her abdomen, above Poupart's ligament, on either side, ribandlike cracks in the skin. These are seldom produced in that part by anything but pregnancy, and must not be confused with the silvery lines often seen, and which own the same cause. This was sufficient to arouse suspicion. But there was another and more important sign. In a virgin, the orifice of the vagina should be partially closed by the hymen. When a woman has sexual intercourse the hymen is not destroyed; it is only lacerated in one or more places; and even in a woman who has had an abortion, the segments of the hymen are still manifest. But after a child at full term the hymen is very much injured; a few bits may remain, but rarely more. In this woman there was no hymen at all, so we may fairly infer that she has given birth to a child of considerable size.

So much for the causation of procidentia; a few words as to the anatomy; I shall only, to-day, give it so far as it is illustrated in the case before us.

And the first point to notice is this—that the disease is called procidentia of the womb, and if you allow the name to guide you as to its anatomy, you will form a most erroneous idea of the disease. In such a disease as that we are discussing there is procidentia of the womb, vagina, bladder, part of the rectum, of the ovaries, of the bowels, of the liver; and probably everything falls down—indeed, as the case of this woman illustrates, the womb is generally the organ that notably refuses to go down, and the disease might be called procidentia of any organ as truly as of the womb.

The common idea is that the whole womb protrudes beyond the vulva. This is, however, not often the case; generally there is tensile elongation of the neck of the womb, the fundus remaining within the pelvis. The uterine probe passes in five inches instead of two and a half, as in the case we have under consideration. The organ, therefore, that chiefly refuses to descend is the womb, and yet it gives the name to the disease. It is of great importance for you to know the ordinary anatomy of this disease. Almost invariably the bladder comes down. It is so closely connected to the uterus that, as the neck descends, it pulls the bladder down with it. The bladder will be found in front of the anterior cervical lip—not invariably so, but I have never seen a case without it. The vagina is inverted. As regards the rectum, it is seldom found down; sometimes a little pouch is formed in it anteriorly, which is of extreme importance in connection with difficulty of defecation. In this woman there was no descent of the rectum.

To the patient the symptoms are most important, and they form a matter to be very carefully considered in connection with doctrines now entertained regarding displacements of the womb. I have no intention here of expressing any opinion regarding uterine displacement doctrines generally, except that procidentia has a most important bearing on these doctrines, which say that the slightest change of position, a little curve, gives rise to the gravest symptoms. Now procidentia is a displacement of the most extreme kind, and what symptoms does it produce? Frequently none at all. The uterus of the woman of whom I am speaking was not only procident, but it was acutely retroflected. Here was a displacement of the most aggravated kind, and yet the patient complained hardly at all of pain; her trouble was that the womb fell outside when she walked; it was the mechanical inconvenience which disgusted her. One has only to look at the woman, to see that she is in blooming health. Most women, however, do suffer greatly from dragging pains in the groins, hips, and thighs, and from difficulties in urinating and defecating.

If the disease be mechanical, so must the treatment be mechanical. You would not treat a dislocation of the shoulder by administering medicines.

The case before us was not one of an aggravated kind. When the woman was lying down the disease was cured. The pressure relations of the abdomen became at once changed. If we knew—some day we may—some method of influencing these pressure relations of the abdomen, we might cure this disease by such means. If fat occur in the anterior wall of the abdomen and not in the omentum, the womb is generally found high up. We do not know yet how to produce fat in this situation, so we must resort to simpler methods.

One of these is a pessary. Certain shapes take fixed points on the walls of the pelvis, and by their aid form a shelf on which the womb rests, and cannot get beyond. Among such is the disc and stem pessary, and the Zwanck instrument.

Another method, and one especially applicable to unmarried women, or after the childbearing period, is to nearly close the orifice through which the womb comes out, not strictly to restore the perinæum, for it may be anatomically entire, but what is termed episioraphy. When the operation is finished, the mouth of the vagina is contracted, and there is no great gaping orifice. So long as it keeps like this, the womb cannot come out. And this operation cures a great many cases.

I may mention the case of a nurse in the Royal Infirmary of Edinburgh. I need hardly say that few occupations could be worse for procidentia than that of a hospital nurse. The operation of episioraphy was performed upon her, she was cured, and she retained her situation for some years; she then married a second time, and had two children. The childbearing destroyed all the renewed perinæum; the womb came down again. She once more became a widow; I operated again, and she is at the present time a nurse in the Royal Infirmary, and has been so for many years, without any procidentia whatever.

A third method is the T bandage with perineal pad, which is very valuable in a case of this kind as an adjuvant. Suppose that the door behind me is open, and I stand in the doorway. I cannot prevent you from crowding out; you will push by me on one side or the other. So it is with the gaping orifice of the vagina; the T bandage will not prevent the womb from forcing its way out on one or other side of it. Episioraphy is equivalent to shutting the door. Then the T bandage acts like the hand placed against the other side of the door, it exerts a force which counteracts the pressure from within, and forbids passage to anything.

Difficulty in curing, or keeping replaced, varies with variations of one condition—namely, the amount of downward pressure of the displaced parts that has to be overcome.

INDEX.

Supræmia, 144
Sarcoma of uterus, 146
Scybalum in rectum, 139
Septicæmia from burst cyst, 162
Serous perimetritis, 51
Shrinking tumors, 134
Sitting, painful, 70
Slough of cyst, 163
Spasmodic dysmenorrhœa, 109
Spondylolisthesis, 77
Spreading of abscess, 62
Stricture of urethra, 89
Systems in medicine, 30

THREATENED abortion, 9
Treatment of dysmenorrhœa, 117
of hæmatocele, 153
of parametritis, 68
of pyonephrosis, 87
Tumors which shrink, 133
Twins, abortion of one of, 10

ULCER, malignant, of body of uterus, 147
Ulceration of body of uterus, 147
of womb, 29

Unilocular cyst, 156
Uræmia and cholæmia, 128
with uterine fibroid, 135
Ureters obstructed by a fibroid, 134
Urethra, stricture of, 89
Urine, retention of, 140
with air and fœces, 141
Uterine hypertrophy in imperfect miscarriage, 12
Uterus, cancer of body of, 138
irritable, 47
ulcer of, 29

VAGINISMUS, 101
in labor, 103
Virginity, absence of, 170
Vomiting of pregnancy, 121

WHITES, 32

ZINC ALUM in uterine cervical catarrh, 34

www.ingramcontent.com/pod-product-compliance
Lightning Source LLC
Chambersburg PA
CBHW021805190326

41518CB00007B/463